SOE IN SCANDINAVIA

The author of this as of other official histories of the Second World War has been given free access to official documents. He alone is responsible for the statements made and the views expressed.

S O E in Scandinavia

CHARLES CRUICKSHANK

Oxford New York

OXFORD UNIVERSITY PRESS

1986

Oxford University Press, Walton Street, Oxford OX2 6DP
Oxford New York Toronto
Delhi Bombay Calcutta Madras Karachi
Kuala Lumpur Singapore Hong Kong Tokyo
Nairobi Dar es Salaam Cape Town
Melbourne Auckland
and associated companies in
Beirut Berlin Ibadan Nicosia

Oxford is a trade mark of Oxford University Press

British Library Cataloguing in Publication Data
Cruickshank, Charles
SOE in Scandinavia.
1. Great Britain. Army. Special Operations Executive
2. World War, 1939–1945—Underground movements—Scandinavia
I. Title
940.54'86'41 D802.S34
ISBN 0–19–215883–X

Library of Congress Cataloging in Publication Data
Cruickshank, Charles Greig.
SOE in Scandinavia.
Bibliography: p. Includes index.
1. Great Britain. Special Operations Executive.
2. World War, 1939–1945—Secret service—Scandinavia.
3. World War, 1939–1945—Secret service—Great Britain.
I. Title.
D810.S7C77 1986 940.54'86'41 85–21501
ISBN 0–19–215883–X

Set by Wyvern Typesetting Ltd.
Printed in Great Britain by
Billing & Sons Ltd., Worcester

Preface

This book describes the part played by the British Special Operations Executive (SOE) in recruiting, arming, training, and dispatching men of the Danish and Norwegian resistance movements during World War II, and its role in co-ordinating the work of these movements with the plans of the Allied high command. It also describes SOE's activities in Sweden. It is not a history of Danish and Norwegian resistance as a whole. Most of the underground attacks on the German occupying forces were carried out independently of SOE; but it is often difficult to distinguish indigenous activities, and those mounted from outside. Many notable achievements belong almost exclusively to SOE's men, although as often as not they had some local help. At the other end of the spectrum there were purely indigenous activities, although as often as not they depended on arms and explosives sent in by SOE. While the HQ staff of SOE was wholly British, though a few Danes and Norwegians were closely associated with it towards the end of the war, the agents sent in from Britain were almost exclusively natives of the country to which they were sent. During the latter part of the occupations the United States Office of Strategic Services (OSS), SOE's American opposite number, was also closely associated with SOE.

I am deeply indebted to the distinguished leaders of the Danish and Norwegian resistance movements, respectively Ole Lippmann and Jens Christian Hauge, and to Christopher Woods, the scholarly custodian of the SOE archive, who have been generous with help and advice.

C. G. C.

July 1985

Contents

MAPS

Abbreviations

AMA	Assistant Military Attaché
ANCC	Anglo-Norwegian Collaboration Committee
BOPA	*Borgerlige Partisaner*
C	Symbol, Head of SIS
CD	Symbol, Head of SOE
COS	Chiefs of Staff
COSSAC	Chief of Staff to the Supreme Allied Commander
D/F	Direction Finding
EH	Electra House
FG	*Feldgendarmerie*
FO IV	*Forsvarets Overkommando* IV
GFP	*Geheime Feld Polizei*
GS(R)	General Staff (Research)
ISSB	Inter-Services Security Board
JBC	Joint Broadcasting Committee
JPS	Joint Planning Staff
LCS	London Controlling Section
MGB	Motor Gunboat
Milorg	Norway's Secret Army
MI R	Military Intelligence, Research
MOI	Ministry of Information
MSC	Motorized Submersible Canoes
MTB	Motor Torpedo Boat
NIC 1	Norwegian Independent Company
NS	Nasjonal Samling

OSS	Office of Strategic Services
PWE	Political Warfare Executive
RNN	Royal Norwegian Navy
RU	Research Unit
SD	*Sicherheitsdienst*
SFHQ	Special Forces Headquarters
SHAEF	Supreme Headquarters, Allied Expeditionary Force
SIS	Secret Intelligence Service
SL	*Sentral Ledelse*
SO 1	Special Operations (propaganda)
SO 2	Special Operations (sabotage)
SOE	Special Operations Executive
USAAF	United States Army Air Force
W/T	Wireless Telegraph

Introduction

In the first months of World War II British clandestine operations against the enemy were in the hands of three bodies created in 1938. Section D in the Secret Intelligence Service (SIS) headed by Major Lawrence Grand, RE, was concerned with 'attacking potential enemies by means other than the operation of military forces', later refined to 'sabotage'. Department EH (so-called from Electra House where it worked with Foreign Office cover) created and disseminated anti-German propaganda, although Section D prepared its own for the countries where it operated. EH was under Sir Campbell Stuart, the Canadian second in command of propaganda against the enemy in World War I. A branch of the War Office, MI R, originally General Staff (Research) (GS(R)), with Colonel J. F. C. Holland, RE, in charge, developed methods of guerrilla warfare.

The Prime Minister (Winston Churchill) approved the amalgamation of these bodies in July 1940 to form the Special Operations Executive (SOE) at first divided into two main branches: SO 1 (propaganda) and SO 2 (sabotage). After prolonged inter-departmental infighting, SO 1 was hived off as a separate agency, the Political Warfare Executive (PWE), in August 1941; but SOE continued to disseminate in occupied countries material supplied by PWE. SOE also shared in the training of propaganda agents for service overseas, and fed to PWE information from the field for propaganda campaigns.

The first ministerial head of SOE was Hugh Dalton (later Lord Dalton), Minister of Economic Warfare. When he became President of the Board of Trade in February 1942, he was succeeded by Lord Selborne, who remained in office to the end of the war. The first executive head (symbol CD) was Sir Frank Nelson, who was appointed in August 1940 and resigned in May 1942 owing to ill health. He was succeeded by his deputy, Sir Charles Hambro, a merchant banker, who resigned after a policy disagreement with Selborne in September 1943. The third and last CD was Major-General C. McV. (later Sir Colin) Gubbins, who had been in charge of operations before becoming Deputy CD.

'Country Sections were the organizational bricks on which SOE's staff pyramid rested.'[1] Lieutenant James Chaworth-Musters, RNVR, who was serving in the British Consulate in Bergen when the Germans invaded Norway, was detailed to form what became SOE's Norwegian Section. In November 1940 it was expanded into a composite Scandinavian Section, with Hambro as Controller, Harry Sporborg (a leading City solicitor) as Head of Section, and Commander Frank Stagg as principal assistant. In 1941 a Sweden and Baltic subsection, and a Danish subsection developed, with George Wiskeman as head *de jure*, and Sporborg *de facto*; and on 1 January 1942 Lieutenant-Colonel J. S. Wilson was brought in from Training Section to head a full-blown Norwegian Section, and for a time the Danish subsection.

The latter became a Country Section in its own right and was taken over by Lieutenant-Commander R. C. Hollingworth. Reginald Spink, and Albert Christensen—the former had been teaching in Denmark, the latter with the British Legation in Copenhagen—were recruited into the Section. In the winter of 1942–3 General Sir James Marshall-Cornwall became Regional Head for Scandinavia, but left after a clash of personalities. For the rest of the war the Danish Country Section was managed by Hollingworth. Wilson became Regional Director for Scandinavia in September 1943.

Section D had an organization in Sweden run by Alfred Rickman which was blown in April 1940, having accomplished little. There were no further subversive activities in Sweden (although the Press Attaché, Peter Tennant, later Sir Peter, did keep in touch with anti-Nazi organizations) until Hambro visited Sweden in the autumn of 1940 to lay the foundations of an SOE mission. It was decided that Tennant should concern himself with Sweden and Germany, and with Denmark, until Ronald Turnbull reached Stockholm in March 1941. Captain Malcolm Munthe, who arrived in Sweden from occupied Norway in July 1940, was made responsible for activities in Norway, with the cover of Assistant Military Attaché (AMA). He was replaced in July 1941 as AMA by Major Andrew Croft, his special operations work being taken over by Lieutenant Hugh Marks. Later Croft gave part of his time to special operations. From November 1941 the Counsellor of the Legation, William Montagu-Pollock, supervised all the work of the Stockholm Mission, with Roger Hinks, late of PWE, as his deputy; and in May 1942 Lieutenant-Colonel George Larden was specially recruited by SOE to replace Montagu-Pollock. In December 1942 Captain Thomas O'Reilly, who had earlier come to Sweden to help with SOE planning, took over from Larden. He was assisted by Wilfred Latham, in charge of

SOE's wireless telegraph services, in which Mrs Ann Waring (who briefly succeeded Lieutenant Marks) played a key part for the whole war. When O'Reilly left at the end of 1943, Turnbull became head of the mission by virtue of his experience, but retained responsibility for Denmark. E. M. Nielsen, who had been personal assistant to Sporborg in the early days of the Scandinavian Section, was brought in to look after activities in Norway. In January 1945 Major Henning Nyberg arrived to act as Liaison Officer with the Norwegian police troops training in Sweden. Lieutenant-Colonel Ram joined the Mission to provide expertise on military planning, against the possibility of military operations in Norway. Some members of SOE's German Section served in Stockholm, but only Major Henry Threlfall became involved in Scandinavian operations.

An Anglo-Norwegian Collaboration Committee (ANCC) was set up in London in February 1942 to keep *Forsvarets Overkommando* IV (FO IV) (the Norwegian high command branch concerned with resistance) more formally in touch with SOE activities in Norway, and the plans of the Allied high command. Hambro was the first chairman, and the British members were Gubbins, Sporborg, and Wilson. FO IV was represented by Major (Professor) Leif Tronstad, Major John Rognes, and Lieutenant-Commander Marstrander. When Hambro became CD in May 1942, Gubbins succeeded him as chairman, and the vacancy left by Gubbins was filled—in March 1943—by Commander George Unger Vetlesen, head of OSS's Norwegian desk. At the end of 1942 Rognes was replaced by Lieutenant-Commander Bjarne Oen; and Marstrander, when he was lost at sea in January 1943, by Captain Thore Horve. In June 1943, when Gubbins succeeded Hambro as CD, Brigadier E. E. Mockler Ferryman took over as chairman. The joint secretaries were Thore Boye, and Commander Stagg, the latter being replaced by V. M. Cannon-Brookes.

It was agreed in 1942 that OSS should establish a mission in London. Fears that in Norway this might prejudice 'the very delicate activities of SOE and SIS' by disturbing their lines of communication proved unjustified. In September 1943 a combined SOE/OSS Section was set up in London; and in January 1944 SOE Country Sections were given joint Anglo-American leadership. Kai Winkelhorn became US chief of the Danish Country Section, and Vetlesen chief of the Norwegian. In August the Norwegians became part of a tripartite organization. On 1 May 1944, in preparation for the invasion of Europe, SOE's north-west Europe directorate, including Norway and Denmark, was integrated with OSS to become Special Forces Headquarters (SFHQ), part of Supreme Headquarters, Allied Expeditionary Force (SHAEF). The integration of FO IV was at first opposed on the ground that SOE and OSS had world-wide responsibilities and that security might be endangered; but after D-Day the Norwegians were brought in. Flemming Juncker, who had done yeoman service in the field in Denmark, and Svend Truelsen who had re-established the intelligence service before being forced to leave the country, joined the Danish Country Section. In January 1944 OSS established the Westfield Mission to work alongside SOE in Stockholm. It was headed by George Brewer, and took over responsibility for all W/T matters. Its main task was to organize the Sepals Mission on the Swedish side of the border to report on German naval and military movements in Norway.*

Finally, 'the Shetland Bus Service', originally set up jointly with SIS, must be mentioned (see Chapter 5). It was opened in 1940 by Major L. H. Mitchell to receive refugees arriving by boat from Norway, and to manage a shuttle service for agents moving between Shetland and the west coast of Norway. Mitchell was joined by Lieutenant David Howarth as his deputy and in the autumn of 1942 his place was taken by Major Arthur Sclater, RM. For virtually the whole war the Shetland Bus played an important part in SOE activities in Norway, first using refugee fishing boats, and later three submarine chasers provided by the US Navy.

* The Westfield Mission was more a political gesture than an operational necessity. OSS also wanted to establish itself in Denmark where yet another factor in the resistance equation would have caused problems. Happily Winkelhorn accepted the advice of Ole Lippmann and persuaded his colleagues in OSS to drop the idea.

1
Techniques

THE minimum qualifications for a secret agent are that he should pass unnoticed in the country where he operates, and speak the language. A white man would be quickly spotted in the Far East, even were he bilingual. An oriental might have a better chance within the populous shelter of a cosmopolitan western capital, but outside it he would be at risk. Some Britons could pass as Danes or Norwegians, but few can speak their language. Before a clandestine organization like the Special Operations Executive (SOE) could start work in Scandinavia, it had to find suitably qualified natives.

The Norwegians stubbornly resisted the German invasion of 9 April 1940, but, with only small French and British expeditions to help them, were forced to capitulate on 9 June. Many, anxious to hit back at the occupying power, availed themselves of the escape route across the North Sea to the Shetland Islands, hoping to serve their country from outside, perhaps by joining a 'Free Norway' army, or the British forces. It was among these men that SOE found first-class candidates for training as agents—organizers, saboteurs, wireless telegraph (W/T) operators, couriers, and propaganda experts. There was no comparable flood of men escaping from Denmark where the government had accepted the German invaders without a struggle; but Danes, many of them seamen, did come to Britain in twos and threes from all over the world, also hoping to join in the fight against Hitler. They, too, provided a pool from which men could be selected by SOE for training as agents.

The first problem when these men had been trained and passed as qualified, usually after a practical exercise in which field conditions were simulated as nearly as practicable, was how to get them back to their country. Geography had an important bearing on the method of infiltration. Thanks to its coastline of 1,250 miles broken by long fjords, Norway was accessible by sea. The first Norwegian agents returned by the route they had escaped over, which became so frequently used that it was known as 'the Shetland Bus Route' (see Chapter 5). It was by no means an easy way back. The North Sea could be a dangerous enemy, and with the passage of time the Germans improved their defences, supplementing their coastal watch with look-out boats. Luck could play an important part. In February 1942 an agent landed from a fishing boat at a point on the coast, eight hours from Trondheim, his ultimate destination, in broad daylight with no one to receive him. He booked a passage on the coastal steamer to Trondheim where a German soldier helped him to carry his baggage ashore—a rucksack and four heavy suitcases. No contact had been arranged for him in the town, but he found his way to the house of a friend. Such easy progress was exceptional.

Agents were, of course, sent into Norway by parachute; but of all the European countries in which the Allied air forces served the resistance movement, Norway presented by far the greatest difficulty. The enemy's defences were augmented by natural hazards. The mountains made low flying dangerous, and bad weather, which could not be accurately forecast, caused many flights to be abandoned. The barren landscape and unreliable maps made navigation difficult.* Icing was another danger. Most flights were to southern Norway and the Swedish border area, calling for round trips of 1,500 miles and maximum fuel loads, with a corresponding restriction in the payload available for agents and stores. It was difficult to find good dropping zones in the mountains. They had to be a safe distance from German anti-aircraft batteries, airfields, and strong points, which meant locating them in remote valleys and plateaux so that stores had to be carried a long distance from the reception point. Reception committees, which had to endure great hardship waiting for aircraft, often suffered in vain. They displayed much ingenuity getting the stores to safety. On one occasion a committee rode the containers down the mountainside like toboggans. It was not unusual for parachutes to fill with wind and drag containers far from the landing point before the ropes tangled and collapsed the parachute. Sometimes German aircraft tried to follow parachute operations, and once contrived to drop their own parachutists to a reception point; but the committee spotted them and made a successful getaway.

Infiltration by air greatly increased towards the end of the war. In 1942 eleven sorties dropped twenty-one men into Norway, but the final figures for the whole war were 717 successful sorties out of 1,241 attempts, which dropped 200 men and nearly 8,500 packages and

* In the latter part of the war, the RAF's *Rebecca/Eureka* homing device, which guided an aircraft to its destination through a signal transmitted from hilltops in Norway (and attics in Denmark) was considered to be more successful in these countries than anywhere else in Europe.

containers. Twenty-three RAF and five United States Army Air Force (USAAF) aircraft were lost. It required great courage to fall from an aircraft, in the dark, perhaps into the arms of the enemy, but few hesitated to take the plunge. A rare exception was recorded by the RAF dispatcher:

> On receiving green light tried to get passenger to jump but at first he refused and made several counter suggestions, all contradictory. He did not want to put his legs down the hole, and showed very little taste for the job. Aircraft flew backwards and forwards for 28 minutes before he was induced to jump on being told the aircraft could not fly about any longer.

The third method of entry into Norway was across the border from Sweden, difficult in the early years, when the Swedish authorities were on the look-out for any infringement of their neutrality that might upset the Germans, but easier with the passage of time and the growing conviction that the Germans could not win the war. The infiltration of agents from Sweden was managed by SOE in Stockholm. A group of men destined for Norway would be installed in a flat in the capital for briefing. They were given return rail tickets to the Swedish holiday resort nearest their point of entry into Norway, so that, if they fell foul of the police, they could claim that they meant to return to Stockholm after their holiday. The group would cross the border on foot by night and make their way to their ultimate destination through a chain of safe houses.

The first SOE agents entered Denmark by parachute, as the sea voyage from Britain was too long and too dangerous; but there was a good deal of clandestine traffic across the narrow stretch of sea separating Denmark and Sweden. An agent might take a chance and travel openly on a scheduled sailing; or he might be smuggled on board in Sweden and ashore in Denmark; or he might travel secretly in an 'illegal' boat. Most UK-trained agents arrived in Denmark, where the terrain was much more welcoming than in Norway, by parachute. Out of 414 RAF sorties, 285 succeeded—more than 10 per cent better than the score in Norway—but SOE made things difficult for itself through complicated arrangements for receiving agents. An elaborate system of W/T signals between reception committees, the chief organizer, and London cost much coding and decoding time, and played into the hands of the German direction finding (D/F) system. If a flight was called off or had to return to base after the BBC code message that it was on its way had been sent, there was no way of telling the reception committee what was happening. Things improved in 1944. Denmark was divided into three regions, Jutland, Fyn, and Zealand, with which SOE communicated direct, eliminating much internal communication; and, if the flight was cancelled at the eleventh hour, the fact was announced in agreed code words in a later BBC news bulletin. Thanks to the easier terrain Danish reception committees could handle much greater quantities of stores than their opposite numbers in Norway. In Zealand as many as seventy-two containers were successfully received in one drop from three aircraft. At the end of the war it was planned to receive 240 from ten aircraft alongside a railway where they were to be collected by a special train manned by the resistance; but the armistice came before this ambitious operation could be carried out. As Denmark's only frontier was with Germany, there was no question of entering on foot.

Agents dropped to a reception committee established their identity through an exchange of passwords, and were then escorted to a safe house to be acclimatized and briefed before joining their organizations. Their arrival was notified to SOE through a pre-arranged announcement in the small ads. in a Danish newspaper taken by the British Legation in Stockholm. As soon as the reception leader heard the BBC message announcing the departure of the aircraft, he sent couriers to round up the rest of his team, who assembled near the dropping zone. Ideally transport was parked in a nearby farm as soon as the flight was requested, and the sooner the aircraft came the less danger the Germans would happen on the vehicles in a spot check. The containers were usually taken to a warehouse rather than a farm, since farms were the first to be searched when the Germans suspected a drop had taken place. The accuracy of the drop was important. In Zealand there could be trouble if containers were dropped even a few hundred yards off target. If they were found by a farmer not in the resistance, he invariably told the police in self defence, which compromised all dropping zones in the neighbourhood.

The differences in the methods of sending agents into Norway and Denmark were paralleled by differences in their way of life within the countries. In the remote districts of Norway, they could live safely for long periods without papers or cover story. 'Alibis' were needed only when they paid a visit to an inhabited area. London-trained agents were given a forged Norwegian identity card, and occasionally a ration book, before they left; but they were for use only in the first few days. Genuine papers had to be acquired as soon as possible. It was difficult to provide perfect forgeries. One agent, dropped near Bergen in December 1944, found his identity card did not bear the local stamp. Worse, it was printed on paper much stiffer than the official Bergen cards. He was lucky in that these discrepancies passed unnoticed—the Germans often tested identity cards by feeling them. The forging of ration books was particularly difficult since colour and format were changed frequently, and it was almost impossible for London to keep up with the changes. Some high grade agents in Norway whose tasks required them to carry a gun, protected themselves with State Police identity cards. Two were arrested in August 1944, but so convincing were their documents that they were released with apologies.

The problem of papers was easier in Denmark in the first years of occupation while Germany still aimed to make the country a model protectorate. The Danish police and the registration and taxation authorities often came to the rescue of resistance men in difficulty over papers. Further, as there was no system of compulsory labour, employment cards were unnecessary. The only paper needed was the *Legitimationskort* (identity card) which had a photograph of the holder, the place and date of his birth, his occupation and address, but was neither numbered nor fingerprinted. Many Danish underground organizations had their own forged-paper sections, usually including a competent photographer. They would either prepare an identity card with all personal details, or leave blanks for the agent to fill in, so that he alone knew his cover name and other particulars. Agents in Denmark often supported their identity card with a hunting or driving licence, letters addressed to the name on the card, tax receipts, and kindred documents. For some sabotage jobs, Germans or high level Danish collaborators were assassinated, their special papers copied, used for a particular operation, and then immediately destroyed.

SOE-trained agents for both countries were provided with cover stories before they left Britain; but, as in the case of forged papers, these were no more than a stopgap. The agent must provide himself with a new cover story in the light of circumstances. At best his cover would get him safely through street and restaurant controls when he would simply have to confirm the information in his identity card. If he was subjected to a full-scale interrogation, he was expected to use his cover story to keep German counter-intelligence from tracking down other members of his group; and, when his captors had proved his story false, to disclose a second equally false story admitting a fictitious connection with the resistance. This gave his fellows more time to go underground.

In Norway faked papers and cover stories were less essential for underground activities than in any other west European country. The large expanses of uncontrolled territory, both on the coast and inland, allowed many agents to dispense with cover apparatus, especially in winter, and to continue operating after their papers and cover stories intended for inhabited localities had been blown.

It was less easy in Denmark with its smaller area and more concentrated population. A variety of cover techniques was used. One was to double the identity of a living person without telling him, either by picking a name from the telephone directory, or through personal knowledge. This had the advantage that it usually secured safe passage through controls, and that the identity card would stand up to a telephone check with the issuing office. On the other hand, it had the serious disadvantage that a visit to the address shown would prove the card false. Another danger was that the man whose identity had been stolen might himself be working for the resistance. One agent who relied on this method found to his astonishment that he was doubling a well-known informer. Another belatedly discovered that the man whose name he had lifted from the telephone book had been dead for some months.

It was much safer to double with his approval the identity of a living person, who would move away from his home and live quietly in another part of town. This gave the agent ready-made information about his identity and antecedents. He knew his double was neither under suspicion, nor working for the resistance, and that, if the police took him to his identity card address, it would prove genuine. This method was widely used in Denmark and worked admirably. Registration offices helped with another method. An official would fill in a blank identity card with the particulars needed by an agent, and then slip it into a pile of cards awaiting stamping and signature. This happened so often in one district that the Germans tumbled to the stratagem and subjected all cards in the district to a special scrutiny. Yet another method was to procure a false birth certificate from a friendly parish priest, and use it to get a genuine identity card and ration book locally. The agent would then move from that parish to the area where he was to operate, which had the advantage that the police, who often tested suspects by questioning them about the area where they were living, accepted that the man was an honest newcomer unfamiliar with the district. An attempt to create cover that was too perfect to be true was to brief the agent before he left Britain on the latest scandals, jokes, gossip, and new catch phrases corresponding to 'Wailing Willie' (the air raid siren in Britain).

In selecting cover employment, certain principles had to be strictly observed. The agent had to be able to talk convincingly about his chosen job. He had to have a contact in the firm where he was supposed to work who would vouch for him. One ploy was for a firm to advertise for commercial travellers. The agent would answer the advertisement, get the job, and be given samples and sales literature. This would enable him to travel the country freely. The most common professions adopted, in addition to commercial travelling, were engineering, teaching, and insurance. 'Assistant' was popular since it was frequently used on genuine identity cards without specifying what the assistant did—usually in a government department or magistrate's court. London showed a lack of local knowledge when they gave a young man the identity of a *Prokurist*, a man who travels widely in Denmark to sign papers on behalf of business firms, and who in real life would have been much older. The *Prokurist* became an agricultural adviser, and later an architect, in both cases with the consent of people in these professions who willingly lent their names.

Members of the resistance who joined in Denmark (rather than being dropped in after training in Britain) had no initial identity problem. They carried on as themselves until they were compromised and had to go underground. One man who was arrested by the Gestapo escaped and transformed himself into a workman 'with a dirty face and cheap spectacles'—which turned out to be a mistake as he found it difficult to meet people of a better class, and could neither hire a taxi nor drive his own car. He solved the problem by becoming a doctor. A blacksmith in north Jutland retained his true identity for the whole war—and his calling was especially useful as it enabled the farmer members of his group to visit him whenever they had anything to discuss.

Disguises were frequently used when agents were under suspicion and had to make a getaway, or were meeting members of the organization whose reliability was doubtful, or after a change of identity. A man could alter his appearance by wearing glasses if he did not normally use them, or vice versa. The type of clothes worn could be changed, a different hair style adopted, a moustache grown or shaved. Occasionally plastic surgery was carried out in England before an agent who had been blown was returned to the field in Denmark.

* * *

The nature of the communications used by the resistance in the two countries was affected both by geography and the differing forms of administration. The postal services in Norway were little used because of the internal censorship. Bulky packages or mail addressed to known suspects were openly censored by the State Police. There was random censorship by the postal authorities at irregular intervals of mail from a given area, street, or individual letter-box. In any case, there was such a long time lag between dispatch and receipt of letters in the country districts, and everywhere in winter, that they afforded a poor method of communication. When for some reason the post had to be used, language was carefully veiled. Some agents mailed letters to a false name which a friendly postal official would recognize and hold for collection by the addressee. But the post was never used outside the Oslo area.

By contrast, the postal services in Denmark were used extensively, both by the resistance movement and the intelligence services. There was no routine internal censorship; and, if it had been introduced, the resistance would have been given due warning by contacts within the post office who as a matter of routine gave a warning when German counter-espionage became interested in a particular individual's mail. Nevertheless, it was understood that letters would be so written that, even if they did fall into German hands, not one word would incriminate either the sender or intended recipient. A number of cover addresses was used. Letters might be addressed to a non-existent patient in a hospital, where the post orderly, put in the picture by a member of the resistance, would segregate the clandestine letters, and pass them to a cut-out (intermediary). This system was used in hotels patronized by Germans. Letters were addressed to fictitious German names, intercepted by the hotel porter, and in due course handed to a cut-out. Poste restante was also used in this way, although it was recognized that, if the Germans suddenly became interested in internal mail, poste restante letters would be the first to engage their attention.

When it was discovered that the German troops were in the habit of writing clandestine love letters to Danish girls, using street kiosks as cover addresses, the resistance adopted this system for their own purposes. A pre-arranged genuine German's name would be written on the back of the envelope, which would be picked up by the kiosk owner, and kept until collected. Householders who had no other connection with the resistance were also used as cover addresses. Letters (which they were told came from illegal newspapers, a less dangerous source than the underground proper) were sent to them with their names slightly misspelt, which they would hold until they were called for. In the unlikely event that the householder became suspect, for example because a pro-Nazi postal official spotted the consistent misspelling, he could give away nothing of value beyond the personal appearance of the man, or woman, who collected the letters.

In Norway the use of the telephone varied from district to district; and in Oslo it was by far the most important method of communicating. One organizer there had several flats and offices at which he could be telephoned at pre-arranged times. Veiled language was used to frustrate attempts at phone tapping. In Trondheim the telephone was used extensively until 1943 when it was banned after a warning by friendly officials that the Germans were now regularly listening in to conversations. In the Rjukan area the phone was freely used as most operators were loyal to the resistance. In those exchanges where the staff were pro-Nazi, engineers installed bypass lines which could be used with impunity.

External communication with Britain was by courier through Sweden, by W/T, by fishing boat, and later by submarine chaser (see Chapter 5). Most written messages from south Norway to Sweden were fed from outlying districts to a courier centre in Oslo, whence they were sent on to Sweden through routes established by the underground military organization (*Milorg* (from *Militaerorganisasjonen*, military organization, see Chapter 8)). In the north messages went over independent routes set up by individual organizers. Wherever possible couriers had some legitimate reason for travelling the route they used, which of course meant using different men for the Norwegian and Swedish legs of the journey. One Oslo organizer was served by three couriers, one of whom travelled from Stockholm to the border, the second crossed the border (the most difficult stage), and the third completed the journey to Oslo.

A report from Stockholm in June 1942 claimed that courier activities were more circumscribed in Sweden than in Norway. Many had been arrested there, any papers and money they carried being confiscated. In this way more than half a million kroner destined for the relief of teachers and the clergy had been lost. There were seven courier routes based on Stockholm: a post route to Oslo; a parcel route for large packages carried by relays of Swedish couriers in Sweden, and Norwegian in Norway; a post and parcel route operated in the same way; a passenger route from Oslo to well into Sweden; a second passenger route from Oslo over which the traveller had to make his own way without the services of a conductor; and finally two post and parcel routes to Trondheim.

Given the nature of the terrain in Denmark, the task of the courier there was even more hazardous. The qualities looked for were intelligence, alertness, and level-headedness; and girls were often preferred to men on the ground that they were less likely to be searched. Messages were either memorized and delivered orally, or written on thin paper and carefully concealed. It was laid down that the courier must never work from agent to agent. He or she would receive the message from a cut-out and deliver it to a cut-out at the other end. Further, the courier must vary the route taken and the meeting-place with cut-outs.

There were two main clandestine routes from Denmark to Sweden: from Zealand, operated by the Danish intelligence service, who placed it at the disposal of selected resistance organizations; and from Jutland operated on behalf of the resistance movement only. At first the latter plied between Danish and Swedish ports, but later, because of the increased vigilance of German patrols and scarcity of fuel, the boats went to meeting-places off Anholt or Laeso, where agents and materials were transhipped to Swedish boats bound for Gothenburg. When it was necessary for a man to use the route from Jutland, he contacted his district leader who got in touch with a specified bicycle shop in the district. The organizer of the route had been a cycle dealer pre-war and had found it easy to enlist the support of his previous customers throughout the region. A special code was used to tell the owner of the cycle shop how many men wanted to get away, and he then telephoned this information to a cycle wholesaler in Aalborg. The route organizer kept in hourly touch with the wholesaler so that he would have up-to-date particulars of the numbers seeking evacuation to enable him to lay on the necessary boats. If the Gestapo had made it too hot for someone to remain in his own town, he was taken to Aalborg right away by an agent sent from the organizer. Otherwise he would lie low until he was summoned to Aalborg, description, password, and recognition signals having been agreed before he set off. He would be kept in a safe house until the time came to sail. One man had a full-time job locating safe houses all over the Aalborg district.

From the safe houses the escapers were taken to the harbour immediately before the boat was due to leave—Frederikshavn, or Aalborg, or one of the smaller harbours at Saeby, Strandby, Skagen, or Grenaa, at all of which fishermen were prepared to receive refugees. If the Gestapo was hot on their heels, the escapers would be carried in an ambulance; otherwise they would take the train from Aalborg to the chosen harbour. Strandby was a special case as the railway did not go as far as the harbour. The man escorting the refugees would cycle into the town but just before reaching it would let the air out of a tyre to justify a visit to the cycle shop. There he would say how many men were following on foot, so that they could be allocated to safe houses in good time. Immediately after dark the escapers would be taken to the boat; and, if they were accompanied by small children, they (the children) would be given an injection to keep them quiet. (On one occasion when a boat sank, two children could not be got out because they were drugged.) Escaping British and American airmen (for whom SOE in Denmark was responsible rather than the recognized escape organization MI 9) were described as 'deaf and dumb' or 'Icelandic' on their identity cards.

Both the escape routes carried arms and sabotage equipment on the return journey; and on the outward trip they carried written material which could not be sent by W/T—operational plans, maps, and long reports.

In both Norway and Denmark W/T communication progressed from an uncertain beginning to something approaching perfection at the end of the war. SOE's assessment after liberation was that the organization in Denmark was better than that in any other occupied country, due to the proximity of Sweden, the efficiency of the Danish underground organizations, and the great expertise of the men who built, repaired, and operated the W/T sets. The W/T operator's job was particularly dangerous. Organizers and sabotage instructors could work more or less to a timetable of their own devising. The saboteur was at risk only while engaged on an operation, or collecting explosives from store, always provided his contacts were security conscious. But the W/T operator was required by the nature of his job to advertise his presence to the enemy every time he transmitted; and the risks became greater as the Germans increased the number and efficiency of their D/F stations, mobile vans, and portable sets—the last used for final pinpointing. In Norway D/F activity varied greatly. It was much more intensive in the south than in the north, and in the south, most intensive in Oslo. In Denmark there was much the same level of activity all over the country, it being easier to cover all regions than it was in Norway. Although the crucial importance of security was recognized right from the beginning, there was a good deal of carelessness, with disastrous results, especially in Denmark; and even towards the end of the war, experienced operators took risks which might have led to serious trouble. One operator was using an attic in the hospital at Roskilde in Zealand when the patients' radios picked up his signals; and, as he was leaving, he saw a detector van and two truck loads of soldiers, who stormed a nearby house. They had mistakenly pinpointed his transmission to this house, probably because an iron bridge had affected their D/F apparatus. This operator had a charmed life, for on another occasion he had to get a vital message through to London and found a platoon of Germans and a D/F van outside the house from which he intended to transmit. He sauntered past them and, thanks to a first-class operator in London, got his message through in record time. As he signed off, his look-out, the lady who lived in the house, saw the Germans approaching. The operator hid the set in the attic, wandered into the garden where he nonchalantly picked a bunch of flowers, and then took his departure at a leisurely pace. The soldiers 'stared at me and the flowers and what conclusions they came to I don't know; but the result was—free passage!' On yet another occasion he found himself transmitting from a house where the Germans staged a training exercise. 'Cannon roared and troops stormed over the fields and into our garden where the battle raged back and forth!'

Strict security was essential for the survival of the W/T operator. In inhabited areas he would never transmit from the same house without a lapse of some weeks, which meant having access to between twenty-five and thirty different houses; and it was vital to have adequate guards at strategic points round the area where the transmission was taking place. One experienced operator who was transmitting at a time when there was increased German D/F activity posted only one guard, who failed to spot the arrival of the enemy. The operator was arrested and brutally tortured, but gave nothing away, and finished up in a concentration camp in Germany. The firm that serviced the D/F cars revealed that the Germans had been after this man for many months, and had identified more than twenty stations from which he had transmitted. 'He was always easy to find as his key-writing [i.e. his touch on the Morse key] was so recognizable.'

An important step towards the defeat of the German direction finders was made in June 1944 when Lorens Duus Hansen, a Danish radio engineer, who had also been the leading underground W/T operator, devised a method of automatic W/T transmission. Messages were now punched on paper tape at leisure and in relative security, away from the W/T station. The tape was then taken to the transmission point and run through at high speed—as much as 80 words per minute compared with about 15 by conventional manual keying—which, of course, greatly increased the amount of information that could be sent and reduced the time available for the enemy to locate a station. It also made the reception of messages easier and more accurate. This system was known as *Badminton*. Another innovation was the use of an S-phone* (developed for communication with aircraft making drops) for conversation between Helsingør in Denmark and Hälsingborg in Sweden (*Minestrone*).

Security checks in telegrams—the failure of which led to SOE's downfall in Holland, where the Germans contrived to operate the sets of a whole string of agents from Britain—were used; and became more sophisticated after the Dutch débâcle. The earliest practice was to introduce a deliberate spelling error at pre-arranged points in a message. However, the telegram reaching the SOE officer's in-tray in London might be mutilated because of careless or inefficient transmission, by atmospheric conditions, or by errors by the decoders or typists at headquarters, so that corruptions in the final typing might mask the deliberate errors. Later double checks were used—a bluff check, which the captured agent would reluctantly reveal to the enemy, and the genuine check which he would introduce into any message he had to transmit under duress. In 1943 an even more sophisticated system was introduced. Sometimes a pre-arranged question and answer were used. The home station would ask: 'Which is the most popular brand of cigarette?' Under duress the agent would answer sensibly. If he was free, he would give a nonsense answer, perhaps 'Chesterfield', the American brand unlikely to be available in large quantities in occupied Europe.

W/T service between Copenhagen and Sweden was supplemented by a photographic communication system. Typewritten messages were photographed on 35 mm film sent undeveloped in specially designed containers, for example, the handle of a shaving brush which, if examined by the police, would spring open and expose the film to light, destroying the message. Later rectal and vaginal containers were used. The advantage of the photographic message was that it could be lengthy,

* The S-phone, developed by SOE, was a portable radio-telephone which allowed ground-to-air conversations over considerable distances. It had a 'silent' microphone which enabled the agent on the ground to speak in a whisper.

did not have to be encoded, and was sent relatively easily and safely. Prints from the negatives developed in Stockholm were sent to London by air.

One of the greatest threats to the resistance groups was posed by informers—in both Denmark and Norway. Without them the enemy counter-espionage system would have had a much harder task, although it was formidable in its own right. In Norway there were the State Police—pro-Nazi young men who had replaced most of the existing State Police soon after occupation; the Criminal Police, whom the Germans had found it impossible to replace as they could not find adequate substitutes, and who were therefore less hostile to the underground than the new State Police; and finally the Special Food Police, who worked against the black market, and were mostly loyal Norwegians, who sometimes found themselves unwittingly clashing with the underground. There were two Norwegian Nazi organizations: Nasjonal Samling (NS), assistant police who were used as informers; and the Hird, a National Socialist youth movement—'hysterical adolescents' who helped to track down patriots 'with malignant efficiency'. The Germans had their own professional counter-espionage bodies, headed by the Gestapo, and supported by the *Sicherheitsdienst* (SD); *Abwehr*; *Geheime Feld Polizei* (GFP); and the *Feldgendarmerie* (FG). Of these bodies, the agents interviewed by SOE after the war 'retained a clear-cut impression only of the Gestapo'.

The informers who worked hand in glove with these organizations came from all levels of society, for example, girl-friends of German soldiers, poor people, susceptible to bribes in money and kind, and 'prisoners' planted in Grini concentration camp. Some believed a German victory was inevitable, and that it would pay them to be on the winning side. By far the most dangerous organization was the Rinnan–Grande group, which operated from Oslo to Nordland. Rinnan, 'a hump-backed and immoderately greasy dwarf possessed the lust for power and self-importance that is so often congenitally the psychological concomitant of the freak'. Together with his partner, Ivar Grande, he carried out sustained and successful attacks on SOE and *Milorg*. In early 1943 they were responsible for the execution of ten young resistance men in Trondheim.

In October 1944 London told Mission *Antrum* in Alesund that Grande had just moved into their area and that his liquidation would be 'greatly appreciated'. Ivar Naes, who had taken part in several SOE operations, was invited to do the job. He borrowed a rowing boat moored off shore near Grande's house and lay in wait for several nights until he learned that, ever since Grande had been threatened by 'a group of loyal but rather rowdy Norwegians', he always returned home in daylight. Naes then planned to break in, armed with hand grenades, but as he crept up to the house Grande let out his dog which barked so furiously that the informer hastily retreated and put out all the lights. Proposals to shoot Grande as he left Gestapo headquarters, using a silent Welrod pistol, were rejected because it was believed he wore protective armour. Finally, on 12 December 1944, he was pursued as he cycled home in the late afternoon by two *Antrum* men in a stolen car, who emptied the contents of a silent Sten into him.

Even boys were prepared to collaborate. A resistance contact in Gestapo headquarters in Oslo intercepted a letter from one offering information about *Milorg* groups in several districts. Two men disguised as State Police told the boy how grateful the Gestapo was for his help and invited him to tell all he knew. He did, and asked for Kr. 100,000, whereupon they shot him. His body was left in the main street and an account of the sordid affair published in the underground press as a warning.

The extreme danger posed by informers in Norway was fully recognized. One man could blow a whole group, and the knock-on effect could lead the Gestapo to groups in other parts of the country. Therefore, at the request of the Norwegian high command, the Norwegian Section selected a team of four agents for the specific purpose of liquidating known traitors. It was recorded of one that he had considerable experience of crime including murder, and had been 'in the field' for thirty years, which was more than could be said of any other member of SOE.

The quartet (Operation *Bittern*) was trained in the use of silent weapons, pistol shooting, the preparation and administration of sleeping draughts and poisons, breaking into buildings, safe cracking, and unarmed combat. The team, dropped on 11 October 1942 (after three unsuccessful sorties), was equipped with an unusual assortment of weapons, including morphia syringes, six bottles of Q poison, nine of U poison, three boxes of ether pads, eight lethal pills, a set of burglar's tools, and handcuffs. One item had to be specially indented for. In the words of the indent, two of them

> have not moved in the past in the best society, but are exactly the type required. In connection with their peacetime activities they were in the habit of taking a

few grains of cocaine in order to key them up. The work on which they will be engaged is of a much more nerve-wracking and important character. They have promised on all their gods (which are different from ours) that the cocaine will be used only in extreme circumstances . . .

The quartet was instructed to go to Oslo and report to *Milorg*. Unfortunately no one in Norway had been told they were coming and a violent storm of protest erupted. The reception committee in the area where they were dropped was furious. 'It was madness to send in men without warning who had not been approved by the Home Front.' 'One is inclined to believe that the people in London whose business it is to arrange these matters possess no intelligence.' People were asking 'if the cooperation between the different institutions in London is really so bad that a lack of judgement on the part of one can destroy the work another has built up with toil and trouble'.

As *Bittern* had been approved by the Norwegian high command, SOE had a clear conscience. In passing the reports on the *Forsvarets Overkommando* IV (FO IV), the branch of the Norwegian high command concerned with the resistance, Wilson wrote: 'I have nothing to say . . . except that the complaints contained therein and the language used is entirely on a par with the experiences SOE has had over a considerable length of time . . . we feel we can only sympathize with you.' A later report from *Milorg* was so much more favourable that SOE suspected FO IV had specially commissioned it. 'The *Bittern* men have made an excellent impression, and we are very glad to have got them.' They had been used as instructors for six weeks before being sent to Stockholm as ordinary refugees, having carried out no operations. *Milorg* did, however, ask that they should be given warning of any future operations of this sort; and they pointed out that the hit list brought by *Bittern* contained the names of many people 'whom we consider ought certainly not to be killed at present'.

SOE continued to have faith in the principal liquidator, and examined with interest a scheme he put forward to liquidate three Norwegians working hand in glove with the Gestapo. He would arrange for three men ostensibly from the Roads Department to dig a trench outside a certain building, to which he would entice the traitors. 'I intend to get the victims down at some time or other during the day and, after I have killed them, leave them lying there until after 10 p.m.' He would then dump the bodies in the trench. 'I should now have the whole of the night available to bury them, and to fill in their grave again. In this way we avoid unnecessary risk in driving and transport.' The three potential victims were blackmarketeers so there would be no difficulty in arranging a rendezvous. The Norwegian Section liked the proposition, but finally turned it down on the ground that *Milorg* would be upset by the reappearance of the liquidator.

The Danish resistance, which faced the same problem, organized special district groups to track down informers. It was not easy to be certain about an individual's guilt. An accusation might stem from animosity, misunderstanding of something a man had said or done, or perhaps hearsay. But once the district committee was satisfied, retribution followed swiftly. The group in Jutland District II was in two sections, one gathering information about informers, the other an action group dealing with those found guilty. The first comprised twelve former CID men, each with his own network of agents in the field. The second had contacts in the post office who sent it mail addressed to people on their suspect list and to Gestapo headquarters to be examined, photographed, resealed, and returned—except that letters denouncing people were retained so that the sender might be tracked down. The Civil Guard, formed after the police had been disbanded to guard property, also helped by reporting anything suspicious, for example, the unexplained presence of a German car outside somebody's house. The section was also in touch through cut-outs with two members of the Gestapo, one the second in command in Aarhus, 'an old fool who drank too much and when drunk would talk too much'. The other, more junior, was an Austrian who had spent his childhood in Denmark and was well-disposed towards the Danes

The action group was also twelve strong, divided into three teams operating separately. They kidnapped suspects against whom there was a strong case, and in the early months took them to a transformer station for interrogation. Later the wife of a senior police official who was in a German concentration camp provided more comfortable accommodation in her villa. When a proven informer had divulged all he knew, including the names of other informers, he 'was taken for a walk and shot'. Twelve were liquidated in this way in the Aarhus district during the occupation—a small number compared with the national total.

The group in Fyn was considered to be the most efficient in the country. Three telephone engineers and three helpers in the main telephone exchange monitored conversations between Gestapo agents twenty-four hours a day. Contacts in the post office secretly examined letters addressed to certain German authorities and to known informers, and resealed the letters after copying them. Special groups were assigned to watch collaboration or association by Danes with Germans. As in Jutland, police who had been forced to go underground co-operated in providing reports on individuals and in shadowing suspects; and another special group was concerned with the liquidation of proven informers, and the examination of their houses and papers. The information collected was card-indexed, and put at the disposal of the sabotage and military organizations to help them in vetting potential recruits. 'The object of the counter-intelligence organization was to find out who were the Danish Gestapo people, and to liquidate them. The Gestapo were powerless with their Danish helpers.'

The underground security rules applied in all countries, although they were not uniformly observed. Summarized, the principal rules were:

Don't share accommodation with another agent.
Don't commit anything to writing.
Don't talk about operations.
Don't ask what is happening in other districts.
Don't go direct to meetings, or home.
Don't get drunk in public.
Don't keep a mistress.
Don't sit in a public vehicle—stand by the entrance.
Don't disclose your identity card name.
Don't recognize other members in public.
Don't undertake more than one type of activity.
Don't go out with another agent, except on operations.
Always have an alibi (i.e. be able to explain your presence).
Use code-names only.
Look for safety signals.
After ringing doorbell, move away and be ready to run.

The rule most frequently broken was never commit anything to writing. Of course some written records were essential. If telegrams were not kept, confusion was bound to arise sooner or later. In Norway record-keeping was reduced to a minimum by sending telegrams and other communications to Sweden in batches for safe keeping. Many leading agents in both Norway and Sweden kept lists of their fellows—sheer inexplicable folly, which in one instance in Denmark led to the biggest disaster in the whole war in Scandinavia. SOE's Liaison Officer in Copenhagen considered that security there left much to be desired. The guide who took him to his first meeting made no attempt to check if they were being followed; there was no provision for a safety signal, for example, a pot plant in a window to be removed if danger threatened; and, when they arrived at the meeting place four gentlemen's bicycles were neatly stacked at the door. His fears were soon justified.

A man in the telephone exchange who had been providing information was betrayed by a colleague. Under torture he disclosed his connection with the chairman of the Copenhagen committee, who was in turn arrested. Under torture he revealed the time and place of the next meeting, so that the whole committee fell into the hands of the Gestapo. Worse followed. The chairman had kept a diary with the names of district leaders, who were arrested and forced to disclose the names of many subordinate leaders. Estimates of the number of arrests ranged from eighty to eight hundred, but whatever the true figure it was a major disaster—which could have been avoided if the chairman had not kept a diary, or had availed himself of his lethal pill.

SOE's Copenhagen Liaison Officer has left a graphic account of his own arrest, interrogation, and escape. He was interrogated in Shell House—Gestapo headquarters in Copenhagen—for eleven days, when the Germans revealed that they knew a good deal about SOE organizations and personalities in Britain. He was on the fifth floor when he happened to look out of the window and spotted three aircraft coming in very low. He divined that Shell House was their target, and in the subsequent chaos ran from the damaged building and made his way to the home of one of his helpers, taking four different trams to shake off any pursuit.

Although he gave nothing away under interrogation, he claimed that it was virtually impossible for anyone to go through a full Gestapo examination without giving something away. The techniques employed by the Germans in Scandinavia followed the pattern established in Europe generally. The agent would be confronted by a group of Nazis sitting round a table. He would be left standing in the middle of the room, ignored by them while they discussed the papers in front of them, speaking loud enough for him to hear the names of other agents with whom they knew he was associated, and the names of staff at London headquarters picked up in earlier interrogations. The agent would then be questioned and his story recorded. A day or two later he would be brought back, to be told his story had been proved false, and would he now tell the truth? The interrogators would elaborate their knowledge of

SOE as a whole, its training schools, and so on, to convince the agent that, if they already knew so much, there was little point in trying to hide anything. If this failed, the Gestapo would resort to torture, for a short time in the first instance, after which their victim would be well treated. If he still remained silent, the torture would be resumed.

In Norway at least Gestapo interrogations were often sadistic orgies rather than clinical examinations. Many reports showed that the interrogators preferred to get drunk before indulging in violence of any kind. 'When you saw the bottles being put on the table, you knew you were for it.' This is confirmed by one case in Trondheim: the agent

was brought in, seated and studiously ignored, while his interrogators proceeded to get drunk on cognac. When they were drunk, three women secretaries came in to take down anything he said. The chair was removed and he was thrown on the floor. The Germans started to beat him with bare fists and to question him. His interrogations always took place at night. For four nights he told nothing. He had no chance to sleep. During the day he had to sit in a very small cell under a glaring light, which also made the air intolerably hot. For the first night no instruments were used; but later there were wooden sticks and the cat-o'-nine-tails. At every interrogation they threatened to shoot him. He was laid on the table and fettered in handcuffs which with every movement of the hand got tighter. The soles of his feet were beaten with truncheons. Drunk Norwegian police also took part in this interrogation. Eventually he told what little he knew.

SOE's final verdict was:

There is absolutely no doubt that in nearly every case the Gestapo were able to extract a certain amount of information by torture. However tough and courageous a man may be, he had no chance of enduring for long the tortures the Germans inflicted, and it was useless to expect him to do so.

The business of resistance could not be conducted without meetings, during which all present, and those members of the organization whose identities were known to them, were at risk. The rules for the calling and conducting of meetings were therefore of great importance. Ideally, one member would decide the date and place of the next meeting, and pass it to a telephone cut-out, giving only his code-name. Other members would phone the cut-out to be told where and when the meeting would be. If anyone knew about it well in advance, there was the danger he would be picked up by the Gestapo (for example, in one of their fishing expeditions where they arrested a large number at random in the hope of getting their hands on one or two members of the resistance) and be made to reveal details of the next meeting. In this sphere as in all others bad security could be compounded by bad luck. In one unfortunate episode an agent deposited a parcel containing a gun in a left-luggage office, where it was found by the Gestapo. The agent had been careful to check that there was nothing written on the wrapping paper, but he missed an invoice number which led the Gestapo to him—and to another agent in the same flat. So far, two breaches of the security rules, one of them inexcusable.

Now bad luck took over. Under very severe torture the first agent disclosed the address of the chief organizer which he knew had been vacated some weeks earlier. When the Gestapo found the house empty they were furious and told the agent that before they asked any more questions he would be given a hundred strokes with a rubber truncheon 'to teach him not to be funny in the future. They wrapped a towel round his face so that his cries would not be heard, and then meted out the punishment.' The agent then said that the next meeting place was Bredgade 47, an address no longer used. But it so happened that at the flat where the meeting was due the key broke in the lock, and the door could not be opened, and the party decided to return to Bredgade 47 for one more meeting; and twelve agents, including the head of the Danish intelligence service, walked into the arms of the Gestapo.

The financial needs of the first agents in Denmark were provided for by currency taken in from Britain, but it was difficult to find enough, and in any case it was risky to send in large quantities of currency by air. As there was no Danish authority outside the country with whom a financial arrangement could be negotiated, funds were raised locally through well-disposed Danes, who were credited with corresponding sums in London which they could transfer after the war. This system was first managed through Per Federspiel (*Husmanden*) who had offered his services to Britain even before the British Legation in Copenhagen left in April 1940; and after he was compromised, briefly by Jonas Colin, and then by Count Adam Moltke (*Porgy*). The funds thus raised were distributed through specially appointed agents in the districts (Operation *Settee*). There was a *cri de cœur* after the German take-over in August 1943 when the general disruption made many agents short of cash; and later, when some W/T operators could not afford to buy even essential clothing, the chief organizer (Flemming Muus) was told that these most valuable members of the organization must be properly looked after.

In Norway, SOE paid UK-trained agents only so long as they came under their jurisdiction. Thereafter they were looked after by the Military Organization. Most indigenous agents refused to accept any payment unless they had been forced to give up their normal employment. The man in charge of forging documents in the Bergen area, a full-time job, was paid Kr. 300 a month. At the other end of the scale, paramilitary 'rankers' in hiding in the Rjukan area received a corporal's pay of Kr. 5 a day. Payments were, of course, made to cover expenses, mostly the costs of travel. The resistance organization also made itself responsible for the support of families and dependants of agents who had to go underground, were imprisoned, or killed.

SOE Stockholm was embarrassed in February 1944 by the large number of Danish resistance men forced to escape to Sweden who claimed they had been promised up to Kr. 1,000 a month. Some even argued that attendance at a single reception qualified them to a life pension from the British government; and if they didn't get it they became disgruntled and indiscreet. It was recorded in July 1944 that many refugees were regarding SOE's office as a bank. 'We are more than glad to help . . . but are disappointed when many refuse to look for work on the ground they are getting a pension from this office.' The payments 'allow people to live comfortably but not luxuriously. They are on a par with salaries paid to Legation staff.'

A detailed account of the equipment supplied to agents, much of it specially designed and manufactured, would require a volume to itself. Plastic explosive, time pencil fuses, and limpet mines were the bread and butter of the agent in Scandinavia. Plastic explosive, twice as effective as TNT for demolition purposes, had the appearance and consistency of putty. It could be carried without danger and moulded into the most suitable shape for the operation in hand. The time pencil contained a glass tube with a corrosive liquid. When the tube was squeezed the glass broke, releasing the liquid which then attacked a steel wire restraining a spring-loaded striker to detonate the charge. This allowed a delay of ten minutes to twenty-four hours, and even longer. Limpets, used to attack shipping, were about the size of a steel helmet and fitted with magnets so that they could be attached to a ship's hull under water. Although these were the most common devices, there were many others, for example the so-called 'tyreburster', which looked like a 2-inch lump of mud and was scattered on roads to enable enemy vehicles to blow themselves up.

Supplies to SOE's Norwegian Section picked at random from the records of the Camouflage Section included:

25 white kid money belts
25 waterproof bags
130 cardboard cartons
1,000 boxes of Norwegian matches
16 toboggans
22 suitcases, 4 briefcases, 7 wallets 'supplied and aged'
clothing and equipment for 46 agents

And for Denmark:

1 ultraviolet bulb for invisible writing
gold tooth covered with porcelain
War Department markings removed from 40 car tyres and tubes
nasal operation, to enable an agent to return to field without fear of recognition

The long lists of medical supplies included anti-dog scent, caffeine tablets, knock-out tablets ('administered in drink, renders the victim unconscious for two hours'), lethal tablets ('contains a lethal dose of prussic acid, in half-inch tablets'), stimulant and sobering tablets ('packets of 4. See directions on packet'), the suicide tablet ('very small size, can be concealed in a ring. Used offensively by dropping in the victim's drink. Acts within a minute or two'), and the special suicide tablet ('Is quite innocuous unless chewed. Can be carried indefinitely secreted in the mouth, and may even be swallowed without ill effect. Will not dissolve in liquid and is therefore of no use for bumping off').

There were of course huge quantities of more conventional items—Sten guns, Bren guns, pistols, grenades, ammunition, everything required by the contemporary infantryman, delivered by parachute in specially designed containers. However, it was not always possible to satisfy the customer. One team bound for Norway found that the only ski boots in store were size 10, whereas one of them took 8. He had to be content with a second-hand pair which disintegrated after a week's use. The restrained comment from the field was: 'Ski running with such footwear has little to recommend it.'

2

Early days

Sabotage

Section D was interested in Scandinavia because of the importance of Swedish iron ore to Germany, already seen as a possible enemy of Britain. German industry took about 8 million tons a year, from mines in the south through Oxelösund, 65 miles from Stockholm; and in the north through the Baltic port of Lulea, and Narvik in Norway. The last was particularly important since Lulea was always closed by ice in winter, and Oxelösund occasionally.

For Sweden Section D recruited Alfred Rickman. Major Lawrence Grand, in charge of Section D, had asked Rickman's father (a barrister's clerk) if he knew of a bright young man looking for a job. Rickman nominated his son, whose export/import business had just failed. In May 1938 Rickman *fils*, who knew nothing of Scandinavia, or its languages, or of iron ore, or of clandestine work, was asked to visit the iron ore regions, posing as a journalist, and unaware that he was working for Section D. After some months he was instructed to write a book on Swedish iron ore which was published by Faber & Faber in August 1939 with a blurb concluding: 'The best technical advice indicated that there was a gap to be filled by this book and that Mr Rickman was the man to fill it.'

Rickman was now put in the picture. His book would give him long-term cover as a secret agent in Scandinavia. He was to set up as an importer of machinery—plausible for an iron ore expert. He was provided with a number of agencies for reputable British metal products to be handled by Jayandeff Ltd, a London company, through its Swedish subsidiary, Skandhamn. For good measure he entered into an agreement with the Warsaw subsidiary of a French manufacturer to sell its dental materials in Scandinavia. He installed himself in a three-roomed office in Stockholm's business quarter and hired a nearby cellar to house the tools of his trade—plastic explosive, detonators, limpet mines, and the rest.

He looked to three groups to provide him with subordinate agents to

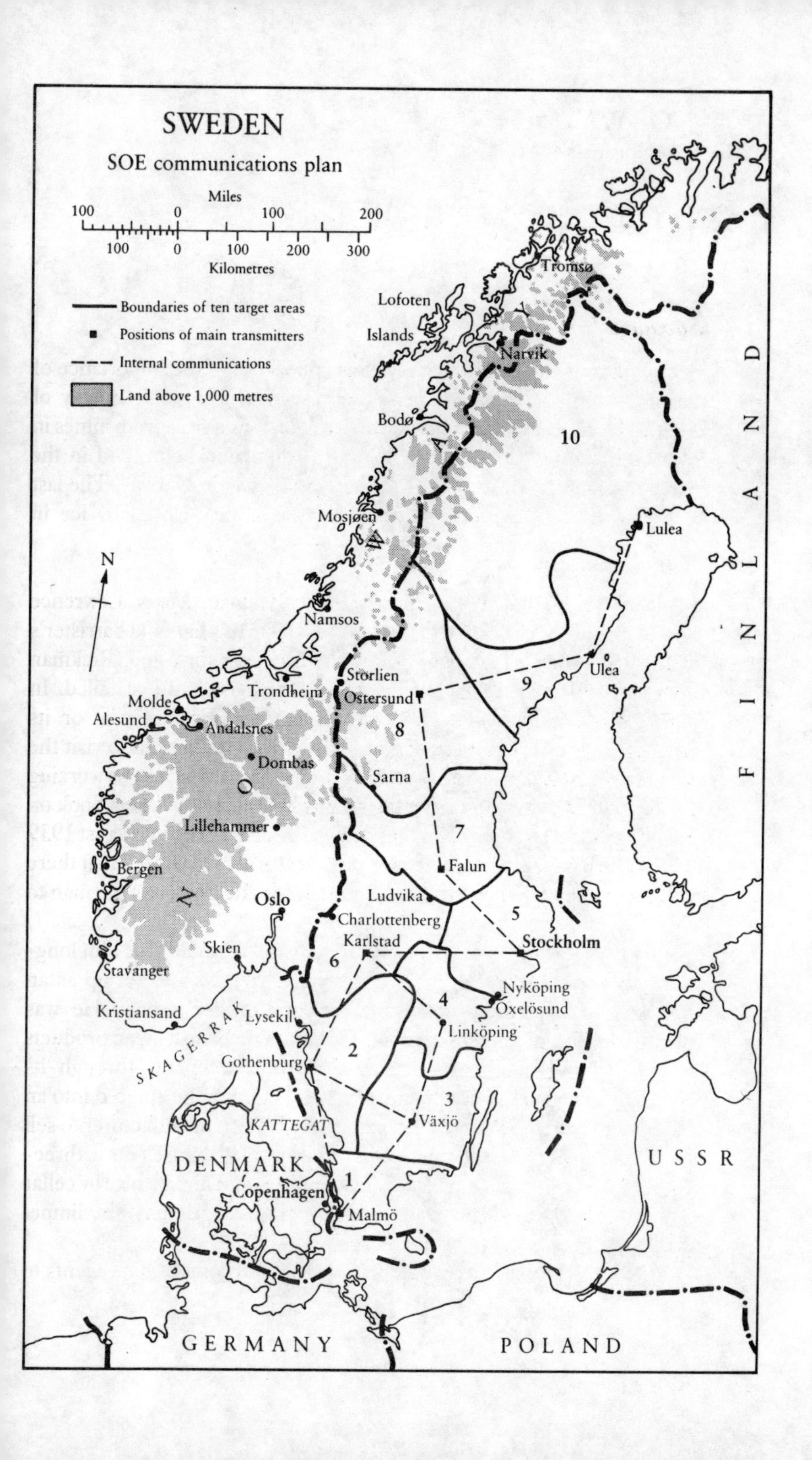
SWEDEN
SOE communications plan
Miles
100
0
100
200
100
0
100
200
300
Kilometres
Boundaries of ten target areas
Positions of main transmitters
Internal communications
Land above 1,000 metres
N
Tromsø
Lofoten
Islands
Narvik
Bodø
10
Mosjøen
Lulea
Namsos
Storlien
Ulea
Trondheim
Ostersund
9
Molde
Alesund
Andalsnes
8
Dombas
Sarna
Lillehammer
7
Bergen
Falun
Oslo
Ludvika
5
Charlottenberg
Karlstad
Stockholm
Skien
Stavanger
6
Nyköping
Oxelösund
4
Kristiansand
Lysekil
Linköping
SKAGERRAK
Gothenburg
2
3
KATTEGAT
Växjö
DENMARK
1
Copenhagen
Malmö
USSR
FINLAND
NORWAY
GERMANY
POLAND

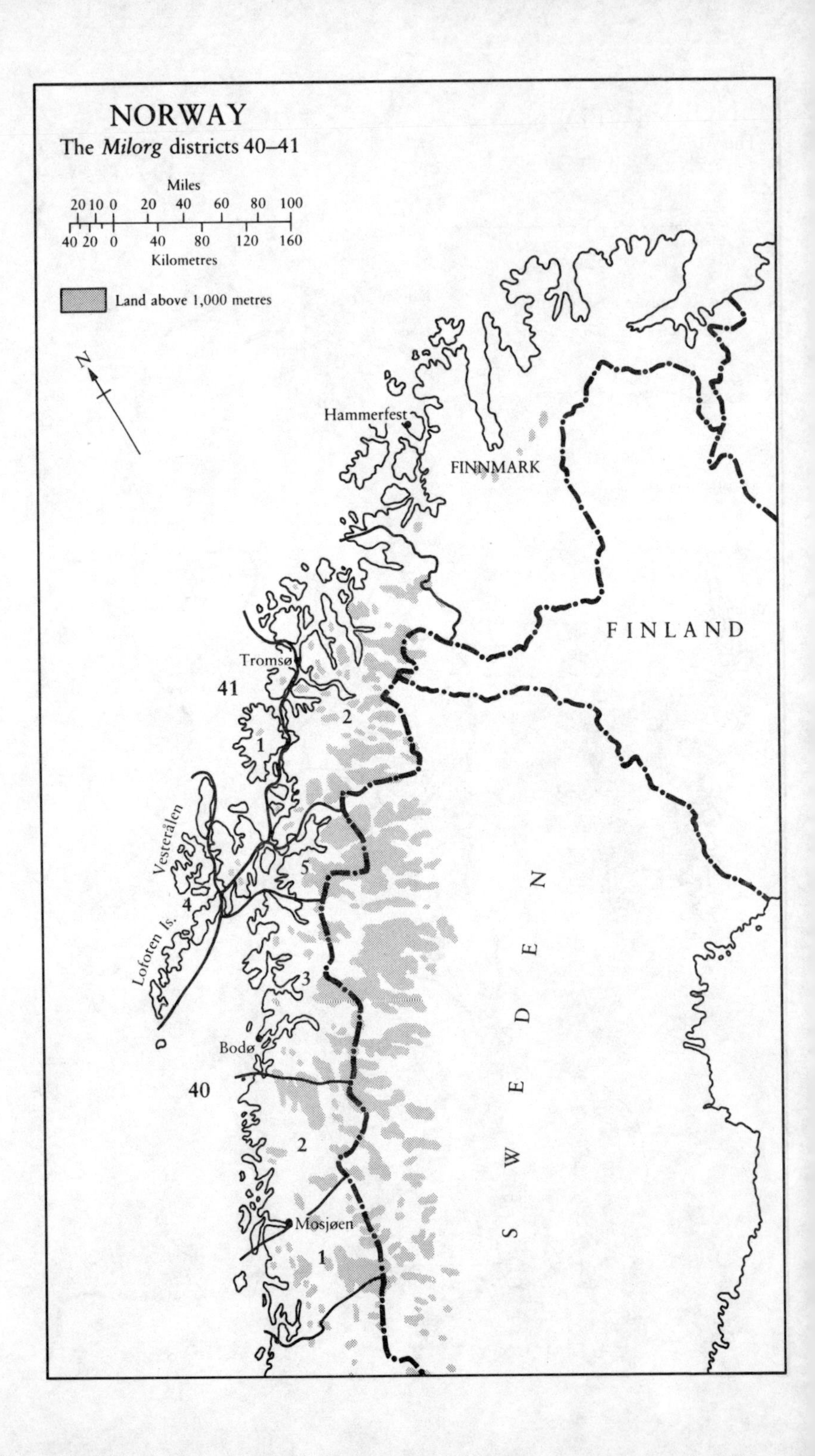
NORWAY
The *Milorg* districts 40–41
Miles
20 10 0 20 40 60 80 100
40 20 0 40 80 120 160
Kilometres
Land above 1,000 metres
N
Hammerfest
FINNMARK
FINLAND
Tromsø
41
2
1
Vesterålen
5
4
Lofoten Is.
3
Bodø
40
2
Mosjøen
1
SWEDEN

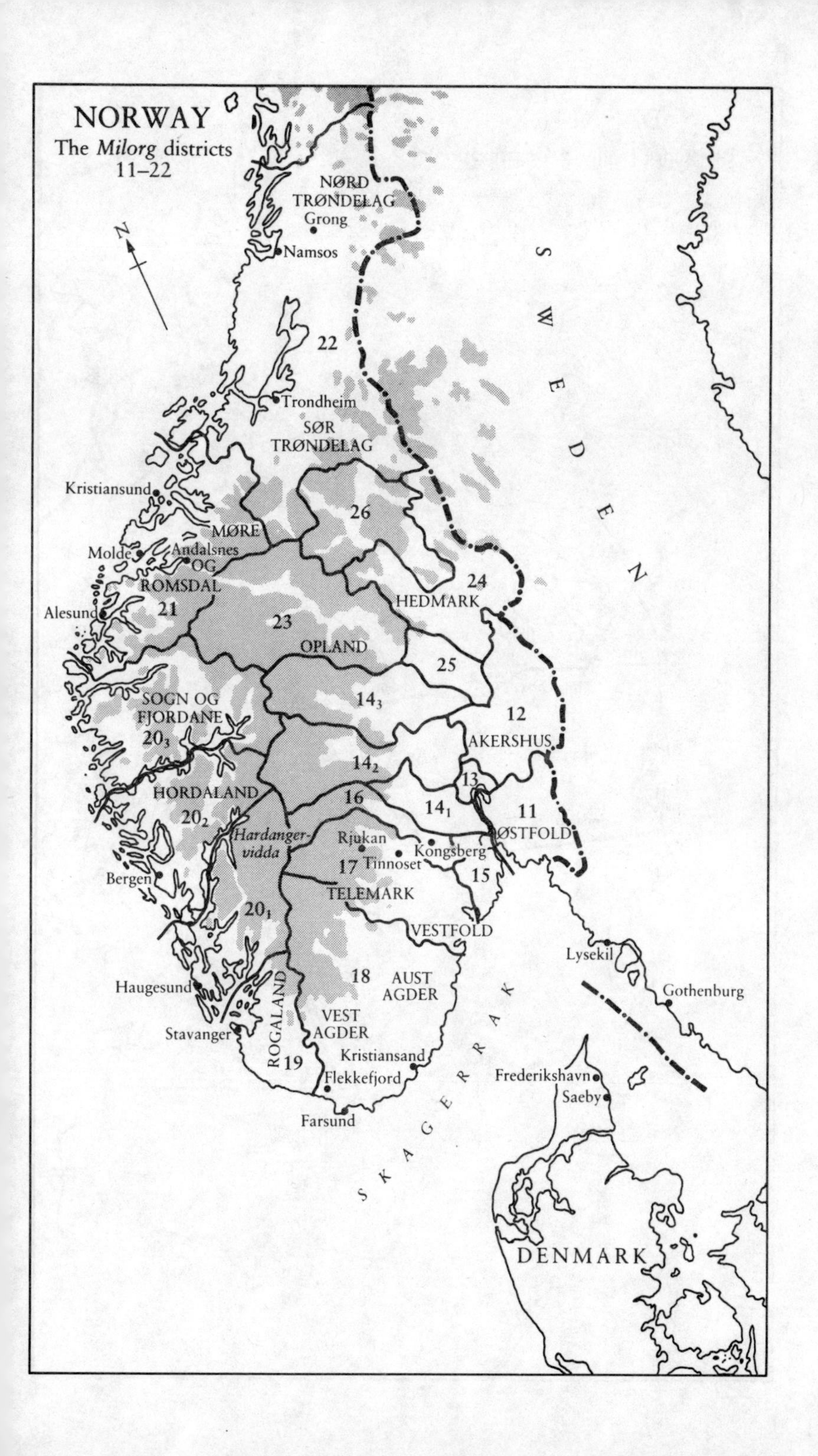
NORWAY
The Milorg districts
11–22
N
NØRD
TRØNDELAG
Grong
Namsos
22
Trondheim
SØR
TRØNDELAG
SWEDEN
Kristiansund
26
MØRE
OG
ROMSDAL
Molde
Andalsnes
24
HEDMARK
Alesund
21
23
OPLAND
25
SOGN OG
FJORDANE
20₃
14₃
12
AKERSHUS
14₂
13
HORDALAND
20₂
16
14₁
11
ØSTFOLD
Hardanger-
vidda
Rjukan
Kongsberg
Tinnoset
17
15
Bergen
TELEMARK
20₁
VESTFOLD
Lysekil
Haugesund
18
AUST
AGDER
Gothenburg
ROGALAND
VEST
AGDER
Stavanger
Kristiansand
19
Flekkefjord
Frederikshavn
Saeby
Farsund
SKAGERRAK
DENMARK

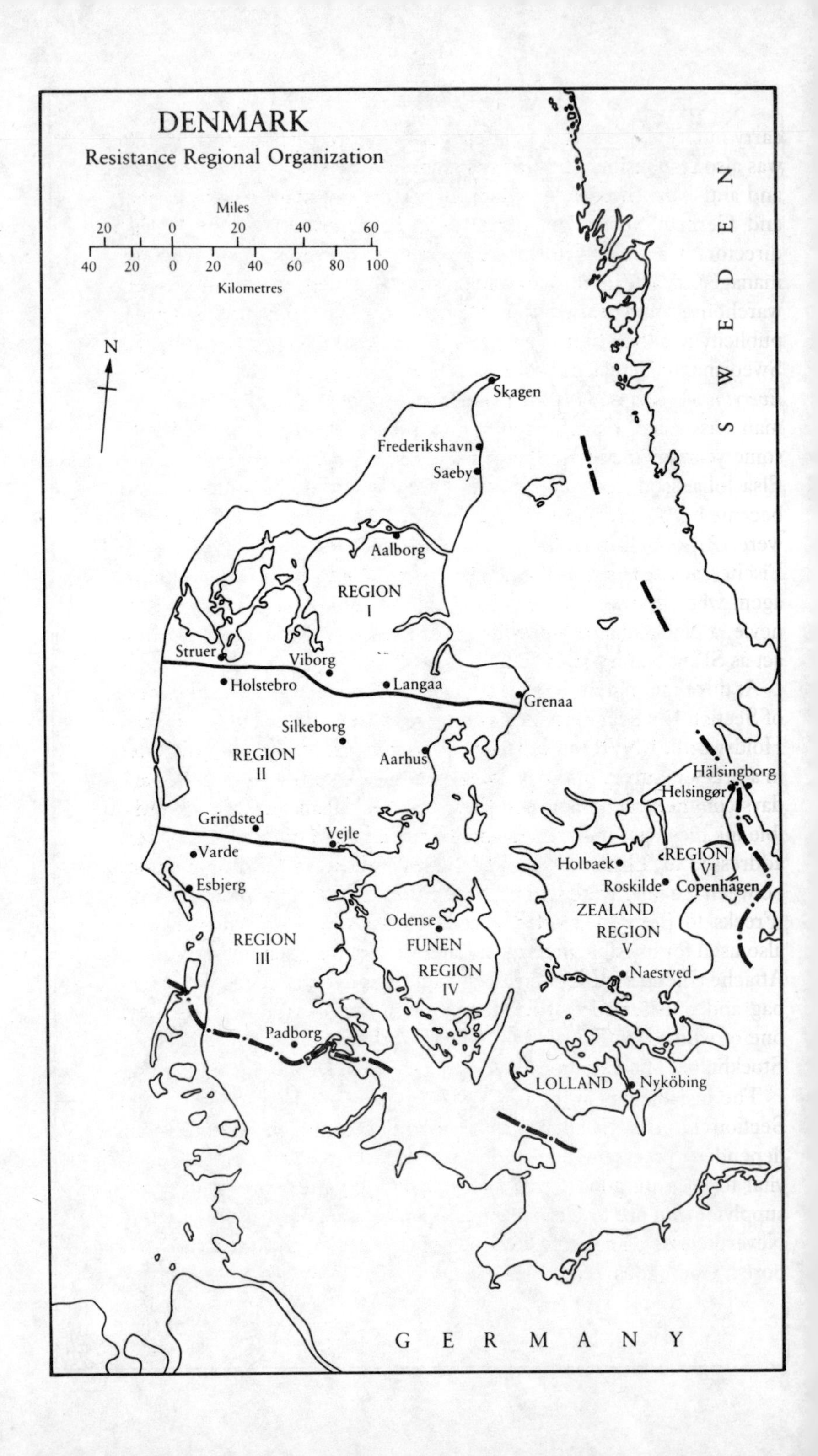
DENMARK
Resistance Regional Organization
Miles
20 0 20 40 60
40 20 0 20 40 60 80 100
Kilometres
N
Skagen
Frederikshavn
Saeby
Aalborg
REGION I
Struer
Viborg
Holstebro
Langaa
Grenaa
Silkeborg
REGION II
Aarhus
Grindsted
Vejle
Varde
Esbjerg
REGION III
Padborg
Odense
FUNEN
REGION IV
Hälsingborg
Helsingør
Holbaek
REGION VI
Roskilde
Copenhagen
ZEALAND
REGION V
Naestved
LOLLAND
Nyköbing
SWEDEN
GERMANY

carry out sabotage and the dissemination of propaganda, for which he was also responsible: the British community in Stockholm; pro-British and anti-Nazi Swedes, of whom there were not many willing to help; and German Social Democrat refugees. Ernest Biggs, the English director of a small firm of tea importers, who was also advertising manager of a leading department store, seemed a neat choice. His warehouse could be used to supplement the cellar; and his experience in publicity qualified him as a propagandist. Ture Nerman, an anti-Nazi Swedish intellectual, publisher of a weekly newspaper *Trots Allt* (*Despite Everything*), was equally well qualified in the propaganda field. Nerman's associate, Kurt Singer, who had been a refugee in Sweden for some years, helped with the printing of propaganda. Another Swede, Elsa Johansson, receptionist at the hotel where Rickman first stayed, became his private secretary, and later his wife. The Social Democrats were: Arno Behrisch, his principal agent in the field; Dr Berman-Fischer, adviser on propaganda; Erich Brost, who wrote leaflets; and an agent who survives only under his code-name of *Dago*. Helmer Bonnevie, a Norwegian businessman, was recruited in November 1939 to act as Skandhamn's representative in Norway.

At this time Ingram Fraser, a former advertising executive, was head of Section D's Scandinavian Section. Second in command was Gerard Holdsworth, RNVR, an experienced yachtsman who had collaborated in a pre-war survey of the Norwegian coast (see pp. 45–6). In the early days communication between Fraser and Rickman was through invisible ink messages in business and private letters. Ordinary letters were addressed to 'Elizabeth' from 'Freckles'. Those with invisible writing between the lines were sent by 'Freddie' to 'Betts'. A third variant, from 'Frecks' to 'Betty', depended on code words. Newspaper wrappers were also used for invisible messages. Later, letters were sent to the Military Attaché (Lieutenant-Colonel Reginald Sutton-Pratt) in the diplomatic bag and transferred to Rickman through 'cut-outs' (intermediaries), one of whom was Harry Gill, the British Petroleum representative in Stockholm. The Legation W/T was also used.

The installations at the iron ore ports were ready-made targets for Section D. But Sweden was a neutral country on whom Britain depended for essential materials, in particular ball-bearings. It was vital to keep on good terms with her and any attempt to reduce the supply of iron ore to Germany might lead to sanctions against Britain. Nevertheless, plans were evolved for sabotaging the three iron ore ports: Operations *Lumps* for Oxelösund; *Sub-Arctic* for Lulea; and

Arctic for Narvik in Norway. (A report by Major J. M. Kirkman of MI 2B in August 1939 said the transfer of ore from rail to ship at Narvik could be disrupted 'by a comparatively simple operation of sabotage'.[1]) *Arctic* contemplated the destruction of the power station supplying the Narvik–Lulea railway, setting up an explosives dump, and sinking British ships alongside the iron ore quays. Nothing was done, however, partly because the Ministry of Shipping failed to respond to a request to provide the ships. Nor was *Sub-Arctic*—to mine the approaches to Lulea, or scuttle a ship in the fairway—followed up.

Lumps did make progress. The possible disruption of Oxelösund had been studied as early as October 1938 but it was not until a year later, after war had broken out, that plans were prepared and exhaustively discussed. On 4 January 1940 Fraser met William Stephenson, the Canadian businessman who became a key figure in Britain's clandestine operations. Stephenson said that Axel Johnson, a leading Swedish shipowner, had been advocating action against exports of iron ore to Germany, and would co-operate, for example, by delaying the arrival of ice-breakers to keep Oxelösund open, or arranging that work on the wharves should be suspended to leave the field clear for saboteurs. Unfortunately he had just been appointed adviser to the Swedish government, which might restrict his usefulness. This proved true. Johnson withdrew his offer to help.

Before any sabotage could be carried out, explosives were needed. In theory they could be brought in in suitably disguised commercial shipments. Small items hidden inside blocks of chocolate; larger in tins of biscuits; and the heaviest things in cases of crude rubber. Dry runs with experimental—and innocuous—consignments through Norway suggested that this method was impracticable. It was then mooted that Norwegian fishermen should pick up supplies at Aberdeen and take them to their home port to await collection, but nothing came of this. The alternative was shipment direct to Stockholm; and by January 1940 several hundred pounds of explosives had been delivered labelled 'military and technical books', in which Swedish Customs happily showed no interest. Sutton-Pratt enjoyed his share in the enterprise, which was kept secret from the Minister, Victor Mallet. The Military Attaché who received the goods, disguised as a French chauffeur, wrote to London: 'We had great fun with the books—real Edgar Wallace stuff, in a dark dirty wood at midnight!'

The First Lord of the Admiralty (Churchill) saw clearly the importance of reducing Germany's imports of iron ore. He wrote in December 1939:

> The effectual stoppage of the Norwegian (*sic*) iron ore supplies to Germany ranks as a major operation of war. No other measure is open to us for many months to come which gives so good a chance of abridging the waste and destruction of the conflict, or perhaps even preventing the vast slaughter which will attend the grapple of the main armies.[2]

But there were obstacles to such an enterprise. 'I see such immense walls of prevention, all built and building, that I wonder whether any plan will have a chance of climbing over them.' The walls included the economic departments of Whitehall, the Joint Planning Committee, moral objections, the neutrals, the War Cabinet itself, the French, and finally the Dominions and their consciences. He was reduced to despair by these 'awful difficulties which our machinery of war conduct presents to positive action'.[3]

On 2 January 1940 the Chiefs of Staff (COS) told the War Cabinet they were in favour of measures to stop exports of iron ore from Sweden to Germany—provided it would decisively shorten the war. When the Prime Minister (Neville Chamberlain) asked the Ministry of Economic Warfare representative if it would, he replied that complete stoppage would be decisive in the long run, but he could not say how long that would be.[4] Perhaps exasperated by this unhelpful advice, Churchill on the same day authorized Section D to go ahead with the sabotage at Oxelösund. Here was a unique opportunity. He might have added Neville Chamberlain to his list of hurdles, for when Section D sought formal approval next day the Prime Minister vetoed the project because of possible damage to Anglo-Swedish relations. At least he agreed to further reconnaissance at the port, which was at once put in hand.

Earlier proposals to sink fully laden ore ships moored alongside, to engineer a go-slow among port and railway workers, and to scare labour away by lacing the ore with explosives, were now rejected in favour of the destruction of the two dockside 'bridge cranes' and the conveyor used to load ore. On 8 January Fraser and Stephenson worked out a new plan. Three weeks later they reported that conditions were ideal. The harbour was still frozen. There were no workmen or watchmen, only military guards at the harbour gates and approaches.

The War Cabinet had not yet agreed to the operation, so Fraser was told to hold his hand and send Stephenson home for consultation. The latter deprecated the postponement, and no longer took the Rickman organization seriously, but was back in London with the new plan on 2 February. The intention was that Bonnevie, Rickman's Norwegian agent, would drive from Stockholm in his Norwegian registered car, which he would park by the roadside, while he spent the night in a hotel. Rickman would take over the car, obscure the number-plates with snow or mud, and drive to Nyköping, a few miles inland from Oxelösund, to pick up Behrisch and *Dago*. While they went on to plant the explosives, he would hide in the woods. Mission accomplished, Rickman would drive the pair to the Malmö–Stockholm railway, and abandon the car, which Bonnevie would then report stolen.

Rickman reported on 9 February that he had visited Oxelösund with Behrisch and *Dago* only to find the harbour working, which ruled out an attack. He offered to try again next day still without War Cabinet approval—but London said no. A few days later he pointed out that the operation might now be more acceptable politically thanks to a round-up of Communists in Sweden. An earlier idea to leave a false trail implicating the Russians might be worth trying? When this was put to the Prime Minister, he again refused permission.

Churchill still hankered after the operation and on 5 March summoned Grand to find out what was happening. It was a sticky meeting. The First Lord said if Oxelösund was to be put out of action it was now or never. Why had nothing been done? Grand replied that the War Cabinet decision against the sabotage still stood. If the project was feasible, Churchill said, why had it not been brought back to the War Cabinet? 'To this neither Colonel Craig nor I made any reply.' After further cross examination Churchill 'went back to the question of why the project had not been raised', to which Grand still had no answer; 'but luckily he went off on to a discussion about Lulea . . .' Churchill took the matter up again with the Prime Minister, and on 8 March the Minister without Portfolio (Lord Hankey) telephoned Section D to say that the all clear had at last been given.

On the same day there was a hitch in Stockholm. The police had earlier stumbled on money paid by Rickman to Behrisch, which he had divided into three packets. One he kept in a book borrowed by a friend was found by the police during the Communist round-up. Behrisch admitted the book was his, but denied knowledge of the packet, which the police had taken to be 'Communist money'. In spite of this encounter, Behrisch and *Dago* were still willing to carry out the sabotage, but then they suddenly changed their minds. If their complicity came to light, it would ruin the Social Democrat image in Sweden, and in Germany the party would be identified with Britain. Later Rickman suggested another reason. The pair were critical of 'our vacillations, incorrectness of information, and instructions to act at short notice' in a project requiring time for preparation and careful attention to detail—a surprising thought since the enterprise had been under consideration for six months. He added that Behrisch and *Dago* would still co-operate 'with the single reservation that they will not personally undertake the supreme risk'. They had tried without success to find substitutes. The only alternative was to use British personnel which headquarters would not allow.

Section D felt badly let down. They had contrived to get approval for the operation on the strength of assurances that 'all preparations had been made and that it was merely necessary to press a button'. The button had been pressed and nothing had happened. The picture changed on 9 April when the Germans invaded Denmark and Norway. Sweden might be their next victim, so Britain could contemplate a bolder line. Without waiting for a visa, Holdsworth flew to Stockholm to take charge. He was unimpressed by what he found. Rickman's organization was too small to be of any use. He must get hold of 'some real toughs who don't mind taking chances'. He had made no plans for swift action in the event of a German invasion. There was only enough petrol in his car for 25 miles. Holdsworth's first task was 'to wangle 1,000 miles out of Shell' to whose manager he had an introduction. He also got in touch with Harry Gill of British Petroleum, who had been acting as a cut-out, the only member of the organization who impressed him: 'a good chap, plenty of guts, knows everybody, speaks fluent Swedish, has a car, and is keen as mustard.' Holdsworth was instructed to destroy the installations at Oxelösund if the Germans invaded, unless the Swedes did it themselves; and to revive the idea of sabotaging Lulea. 'It is necessary to have a really good pile-up of ships blocking the channel for three months.' He was also told to find two safe houses; and, longer term, to organize Norwegian and Swedish resistance movements.

Rickman reported to Section D that he was ready to carry out the sabotage. The reply came the same day: 'Go ahead! Good luck!' But when he met Holdsworth on 13 April they agreed that a new military guard at Oxelösund ruled out an attack. Moreover, since the Germans were expected any day, the explosives must be dispersed. One-third would be left in Stockholm. Another third, limpet mines, would go to south coast ports for use against German ships. The rest would be taken to the country. Holdsworth sent Elsa Johansson to Ludvika, 100 miles north-west of Stockholm, to rent a small house. Later, another would be found near the Norwegian border 'so that when we get Trondheim we can use this new place as a smuggling hide-out for passing supplies back into Sweden'.

Holdsworth planned to lie low in Rickman's flat, but when he went there found he had the wrong key. He thought the flat was being watched and went to a hotel to warn Rickman, who 'ragged me for seeing things'. The plan now was that Rickman, Biggs, Gill and Elsa Johansson should take the explosives to Ludvika—two suitcases-full each. They were to leave on 18 April but deferred their journey for two days—why is not clear. Rickman was to collect Gill's two suitcases on 19 April—it being agreed that Gill should sever his connection with the group—and when he failed to appear Gill telephoned him. Elsa Johansson, already in the hands of the police, answered, and contrived to hint at serious trouble. He at once went to Holdsworth, who arranged sanctuary in the Legation for the incriminating suitcases. Rickman and Biggs had also been arrested, caught red-handed loading the explosives into Rickman's car.

The group were tried in camera, but when the verdict was announced a full summary of the prosecution case was given in open court—most unusual. Rickman had been paid by a foreign power to damage installations at Oxelösund to prevent the export of iron ore. The sabotage had been planned by him, Biggs, Behrisch, and Miss Johansson. Rickman was paid Kr. 35,000 by the foreign power, of which Kr. 10,000 had gone to Behrisch. They had twice visited Oxelösund to destroy the cranes but had abandoned the attempts. Rickman was sentenced to eight years' hard labour for receiving gifts from a foreign power to injure the state, for storing explosives with intent to damage property, and for possessing firearms and ammunition. (He was released on medical grounds after four years.) Biggs was given five years, later reduced to one. Elsa Johansson and Behrisch each got three and a half years. The female governor of the prison where Miss Johansson served her sentence was a Nazi sympathizer 'which added immeasurably to all she was forced to endure'. Bonnevie was arrested by the Gestapo in Oslo in August 1940 because of his involvement with Rickman, and in May 1942 was deported to Germany where he was convicted by the People's Court in Berlin for 'efforts at high treason and treachery', and fined.

The destruction of the Rickman organization came as a shock to the

British Minister in Stockholm, who had been only vaguely aware of what was going on. Holdsworth was left in an awkward position. The police might have spotted his association with Rickman. Or a member of the group might reveal the connection. Lacking a plausible reason for being in Sweden, he hit on an audacious idea. On 14 April he wrote to Grand: 'I want to be appointed by open wire from England, signed by a big name, to the position of head cook, bottle-washer, and general administrator of the Norwegian Relief Fund.' This was a stroke of genius. It would give him freedom of movement in Norway and Sweden, an office and staff in both countries, and 'contact with every Tom, Dick, and Harry—high up and low down'. Grand must arrange for 'Lord Tomnobody' to send a telegram to the Norwegian Prime Minister, then in Stockholm, announcing the Fund. 'I can't enumerate all the things I can visualize around this idea—there just isn't time—but please try to agree to it and I'll promise you results.'

Grand agreed without hesitation; but it took time to get the idea generally accepted. The ever-cautious Foreign Office, unaware of Holdsworth's predicament, was nervous about unfavourable Swedish reactions. Meantime Holdsworth bombarded Grand with requests for speed. 'You must be tired of my reiterating that the Fund must start and start damned quick. It must, Master, otherwise humble servant is in most peculiar position.'

On the assumption that the Fund would be approved, Holdsworth and Gill worked together on a plan. Gill, as a leading member of the British community, was an obvious choice to help. If Holdsworth's connection with Rickman had to be explained away, Gill would say he had brought them together since Rickman's business experience would help to raise money for the Fund. The need for explanations arrived sooner than expected. Holdsworth was in his hotel with Arke Bratt, a Norwegian friend, discussing the Fund, when

> three thugs walked in and asked for me—told me I was under arrest. I asked what for, and they wouldn't say. I asked where their warrant was: the leader showed me a police badge. I was not allowed to phone anyone—had to go right along with the underlings. The chief thug remained behind—took names and addresses and searched all my things, papers etc. There was nothing incriminating in my room. On the way to the police station in a taxi I had the idea they were Gestapo men, and asked to see their badges again . . .

Inside the police station 'I made a first-class fuss and demanded that I be allowed to speak to Baron Beck Fries.* I was refused. No passport, but I had my courier's pass.' Eventually he was allowed to leave, with apologies. He thought his name dropping had saved him, but later decided it was his courier's pass.

When Bratt told Karl Hambro about the arrest, the Norwegian Prime Minister, unaware that Holdsworth had been released, went to see the Foreign Minister and Chief of Police. 'Screams the place down, and even accuses them of doing it under pressure from the Germans, so that even the civil Norway shan't get relief. Then he goes to see Mallet and asks him what he's sitting about for when the General Secretary of Norwegian Relief is languishing in gaol.' This led to a curious meeting with Mallet who summoned Holdsworth to complain that he was trying to do by underground means what the Legation was doing most successfully overground. He suggested that the Relief Fund was a figment of Holdsworth's imagination and that secret agents did not run crying to Ministers of the country where they were operating. Holdsworth replied that the Fund was genuine, and that he had nothing to do with the representations to the Foreign Minister. Mallet cooled down 'and the meeting ended very cordially'.

There was still no confirmation from London that the Fund had been approved. Holdsworth kept in touch with Gill by telephone and when one day he failed to reply it looked as if he had been arrested. When this was confirmed it was obvious that Holdsworth's turn would come next. After discussion at the Legation, it was decided he should get out of Sweden. 'Went back to the hotel to think things over. Warned by girl in cigarette kiosk that the police were up in my room. Didn't go up to interview them. Decided I'd have a stab at Finland. Got visa and steamship ticket and subsequently set sail, leaving my gear behind.'

On 2 May, shortly after Holdsworth's precipitate departure, a letter appeared in *The Times* over the signatures of the Foreign Secretary, the Norwegian Minister in London, the Archbishops of Canterbury and York, the Lord Mayor of London, the Lord Provost of Edinburgh, and Lord Derby, announcing the Norwegian Relief Fund, to be launched at the Mansion House on 17 May, Norway's National Day.

Norway has become a battlefield. Its towns and countrysides are suffering all the terrors of war . . . We appeal to one and all to assist in the work of bringing relief

* The Swedish Foreign Minister, whom Holdsworth had met in connection with the Norwegian Relief Fund.

to Norway and to provide funds to supply food, clothing, medical comforts and necessities to the suffering people . . .[5]

Holdsworth's proposal to return to enjoy the cover afforded by his Fund was greeted with horror. In prison Gill steadfastly denied any connection with Rickman, but the police knew all about his suitcases and proposed trip to Ludvika. He said he had no idea what was in the suitcases and, thanks to his refusal to be browbeaten, was found guilty only of possessing explosives, and fined a mere £15.

Sooner or later Rickman's makeshift organization had to fall apart. Two things hastened its end. Biggs had been under surveillance by the Swedish police from early in 1940 because of his association with the anti-Nazi Ture Nerman. He knew of their interest in him and even referred to it publicly, as something to be proud of. A police inspector, well-disposed to the Allies, warned Martin, the SIS representative in Stockholm, that Biggs was liable to be arrested. Martin told Rickman, who did not take it very seriously. He did stop Biggs working on anti-German propaganda but continued to allow him to visit the Legation and to meet Legation staff socially. This astonished the policeman who later told Martin he could not understand why Biggs had not been completely dropped. He was being shadowed 'night and day' and it was through him that the police had got on to the others.

Martin had not told Rickman that the source of the leaks about his organization was Kurt Singer, Ture Nerman's partner whom Rickman had employed to print propaganda. Martin had known this for eighteen months, and if he had revealed to Rickman that Singer was a police informer surely he (Rickman) would have had nothing more to do with him. Any doubt about Singer's implication disappeared when the police became aware that Nerman and Biggs had met a 'Mr Foster' at the Legation. 'Mr Foster' was the alias used by Fraser when he met Nerman, so whoever passed on information about the meeting believed 'Mr Foster' had been present. Nerman had mentioned 'Foster' to Singer, who told the inspector, who told Martin.

It was not until the end of March, when Rickman's group was hopelessly compromised, that Martin came clean. He told Rickman he had concrete evidence that Singer was an informer who had enabled the police to establish a connection between Biggs, Nerman, 'Mr Foster', and the Legation. He did not admit he had known all about Singer for eighteen months, but implied it was a recent discovery. Why did he keep his colleague in the dark for so long, allowing him to play with fire?

Probably because, if Rickman had known the truth, he would have severed all connection with Singer. The inspector would then have guessed this was a result of the information he had passed to Martin, and that Martin was therefore part of the Rickman group. Even if he were not himself compromised, Martin would have lost a valuable source of intelligence. This is confirmed by Martin's letter of 28 March to Rickman: 'Please excuse me writing in a very dictatorial manner but it is absolutely essential that the following steps must be taken if we are to prevent a very serious blow-up.' Biggs must stop visiting the Legation—which he would have done earlier if Martin had played fair. He must stop noising it abroad that he was suspected by the police. Above all, Rickman must keep to himself the fact he knew Singer was working for the police.

Even if Martin had not followed this selfish and disastrous line, Rickman would have been independently blown by one of the men he tried unsuccessfully to recruit. In December 1939 he consulted Emanuel Birnbaum, a German Jew whose name had been given him by a German refugee in London, about the propaganda he was putting out for German consumption. Birnbaum said it was not up to much, and gave him the name of a journalist, Erich Brost, who might be willing to help. In March 1940 Rickman again saw Birnbaum and asked him to join his organization, when he replied that, although he was against the Nazis, he had no intention of becoming a traitor. He proved his loyalty to the fatherland when he wrote a letter to a friend in Germany enclosing a blank sheet that caught the eye of the Swedish censorship. The police tested it for invisible ink to reveal an account of how Birnbaum had wormed his way into the Rickman organization to learn the names of the leading members, including the top man—Wilson—'whose acquaintance I could not make'. This discovery, made on 13 April 1940, a week before Rickman was arrested, was in itself enough to make thc case against him; and that the invisible message was used by the police is established by the fact that they arrested an innocent man named Wilson, and 'gave him the works, but let him go with a complete apology'.

According to the inspector Rickman divulged virtually everything under interrogation, including the symbols by which he and his colleagues were known. One consequence of the débâcle was that the inspector concluded that British secret agents were all rank amateurs and that he would be unwise to confide in them again. Another was that the Legation telephones were now regularly tapped.

Almost everybody concerned contributed to this inglorious episode. By blowing hot and cold about the proposed sabotage, Ministers in London put the Rickman organization in a difficult position. ''Twere well it were done quickly'—or not at all. If Churchill's assessment was right, and iron ore was of crucial importance, Rickman should have been instructed to build an organization to attack all three ports. Biggs went out of his way to attract attention to himself. *Dago* and Behrisch opted out at the critical moment. If Rickman had set out to give the game away, he could hardly have done better. He broke the first rule of clandestine work—keep agents ignorant of the identity of their fellows. He did not believe he was under surveillance—which Holdsworth spotted right away. Sutton-Pratt thought he had been 'criminally careless'. Britain's first clandestine operation in World War II could not have been worse handled or ended more disastrously.

When Holdsworth came to Sweden on 13 April 1940 he was under instructions to carry out acts of sabotage against the Germans in Norway; to co-operate with the Norwegians in establishing arms dumps in areas likely to be held by the enemy; and to start building a resistance movement in Norway. He had carried out a preliminary reconnaissance of the country in September 1939, when he investigated the canned fish industry ostensibly on behalf of a British importer. He found that the fish canners were worried by shortages of tomato purée and olive oil for packing fish, and of tin plate for the manufacture of cans. If supplies could be increased, it would create goodwill and more important provide a means of smuggling in explosives. He put the idea to Nielsen Moe of the Olaf Preserving Company who agreed to provide a warehouse at Helgeroen in Oslo Fjord where the tins containing explosives could be segregated to await collection.

In September 1939 an office was found—on the eighth floor of a modern block in Oslo—for Helmer Bonnevie, the 28-year-old Norwegian chosen by Rickman to be his agent in Norway. He was advertising manager for the German Telefunken firm, which job he found distasteful, and was glad to have the chance of helping the Allies. He had just opened a small factory to make radio dials and had many business contacts in Norway. It was logical for him to represent the Skandhamn company and he was given a full-time job as 'commercial and secret agent' at Kr. 500 a month. He bought a second-hand car to which he fitted an extra petrol tank. A hidden compartment was built into the front seat on the pretext that the owner carried valuable jewellery samples. One advantage of a Norwegian-based car was that petrol, rationed in Sweden, was freely available in Norway. A Norwegian-born naturalized British subject—Jakob Lund—was recruited to act as intermediary between the Norwegian and Swedish offices.

The orderly development of Section D activities in Norway was, of course, upset by the German invasion. Holdsworth was now expected to work miracles in a matter of days. He was sent elaborate instructions on 14 April, the day after he arrived in Sweden. British troops would land on the west coast of Norway within a fortnight. He must, therefore, find a harbour connected with Norwegian military headquarters by a good road not threatened by the Germans; and create an organization to work behind enemy lines. He should prepare to commandeer transport, find interpreters, provide intelligence, print leaflets, and distribute explosives. As if this was not enough, he must buy a medium wave transmitter to operate a Freedom Radio Station:

> In view of the somewhat restricted road system and the inevitable bombing which will follow the Station being found by the enemy, the Station should if possible be able to be transported by lorry or car and by pack—since it is quite certain that if such a pack set were available it would not be found easily, operating as it would in the hills.

It should be located on the Norwegain side of the border near Sarna where there was a good rail connection. Holdsworth replied that, if he bought a transmitter locally, he would certainly be arrested. He already had operators standing by. All this was pie in the sky. The Germans were much too securely established for anyone to accomplish anything single-handed, which Holdsworth was after 20 April.

Immediately after the collapse of the Rickman organization, sabotage in Scandinavia could be carried out only from Britain. An operation was quickly mounted—the last before Section D became 'Special Operations'. The 11-ton Norwegian fishing boat *Lady* was renamed *V 2 S* and fitted with eight bunks in the fish hold. Plastic explosive, hand grenades, machine guns, carbines and automatic pistols were taken on board; and on 29 May *V 2 S* sailed from Lerwick. Karsten Wang was skipper, Simon Field first mate, and Karl Kronberg engineer, with twelve others making an all-Norwegian crew. Their objective was the district between Sogne and Hardanger Fjords, where they planned to destroy telegraph lines, underwater telephone cables, road bridges, tunnels on the Bergen railway, and factories and power stations—an ambitious programme.

Two days later they arrived at Stenso from which Kronberg and Rubin Langmoe set off to reconnoitre the target area and set up dumps to be used later. Three men they took with them lost their nerve and had to be sent back. Kronberg and Langmoe carried on to Torskevdal where they fell in with an old soldier who gave them a horse and cart to carry their explosives, which they cached at key points. Having made a circular tour under the noses of the Germans, they returned to base on 10 June.

Their operational trip began five days later, their purpose being to plant as many explosive charges as possible within the time allowed by the longest delay fuse—twenty hours—so that they could be safely back on *V 2 S* before the first explosion alerted the enemy. They began with the hydroelectric plant at Alvik where they laid explosives in two of the three water-intake pipes. Then in succession they dealt with telephone pylons, a cutting at a strategic point on the Bergen–Norheimsund road, a road bridge, high tension power lines from Froland and Osteroy at the point where they joined *en route* for Bergen. By the time the pair reached Samnes their twenty-hour time limit had almost expired. They had to abandon their final objective—the railway at Dalseid—and hurry back to base.

The *Lady* had meantime undergone another change of name. She was now the *Hospiz*. Field, accompanied by two others, had gone to Bergen by coastal steamer on 3 June, returning on the ninth. He had contrived to get permission to trade with the *Hospiz* on the west coast as far north as Hardaland, a potentially valuable concession which in the event was not exploited. The remarkable success of the expedition was confirmed by eye-witnesses. The water pipes at Alvik were seriously damaged and the plant was out of action for two months. Most of the other targets were demolished and six well-stocked dumps were established for future use.

A second operation in 1940 ended in disaster. Konrad Lindberg and Frithof Pedersen, who had arrived in Aberdeen from Norway in August, returned by boat to establish W/T communication with Britain. They were safely landed but no more was heard of them. It was later found out that their W/T set had proved faulty and they could not get in touch with the home station. They had been arrested soon after landing and were executed for espionage on 11 August 1941.

One other Section D initiative must be recorded. In 1939 a group of yachtsmen explored long stretches of the coastline of Europe to provide information for use in the event of war. Lieutenant Frank Carr, RN, arranged for seventeen members of the Royal Cruising Club, who could cruise freely without arousing suspicion, to be equipped with Admiralty charts, pilots' directions, and large-scale coastal maps. They would correct and amplify these documents, paying particular attention to road and rail communications that would be important for troop movements. Their survey began at the Franco-Belgian frontier, covered both shores of the Kattegat and Skagerrak, and the coast of Norway as far as Trondheim. Their findings were passed to Carr for collation and transmission to Section D. This work was complemented by an examination of possible landing places by Holdsworth and his fiancée, Mary Thomson, in the course of a walking tour from Stavanger to Bergen.

Propaganda

Although Department EH became fully operational in the spring of 1939 it was not in sole charge of propaganda. A Section D memorandum in April commented on 'the current propaganda chaos' caused by the multiplicity of interested bodies. In addition to EH there was an organization studying propaganda at home and in neutral countries, the BBC's news and features programmes, and the British Council's cultural propaganda. The memorandum did not mention the infant Ministry of Information (MOI) which was preparing to enter the fray. Section D advocated a Policy Committee to co-ordinate all propaganda activities through two bodies, one aiming at the enemy, the other at neutrals.

Section D had itself come early into the propaganda field. In October 1938 it set up a 'Joint Broadcasting Committee' (JBC) with studios at 71 Chester Square and Rocks Farm, near Crowborough, the home of its secretary, Miss Hilda Matheson, to supplement BBC broadcasts by using commercial stations overseas, including Radio Luxemburg. It was not very successful and soon changed itself into a private 'goodwill organization' to provide programmes—scripts and recordings—to publicly owned broadcasting systems. Contributions were made by experienced broadcasters, including Guy Burgess, a member of the Committee who had been with the BBC Talks Department for three years. An offshoot of JBC, United Correspondents, provided feature articles for the press world-wide. These bodies had some success in all four Scandinavian countries, although their work was interrupted by the German invasion of Denmark and Norway.

It was intended that when war broke out JBC should become part of MOI, which it duly did; but wearing its Section D hat it had begun clandestine distribution of its recordings, and was reluctant to stop. In addition to £2,000 a month from MOI, JBC was paid £800 towards the cost of its secret work by Section D, but contrived to charge a large part of its secret costs to MOI. In August 1940 the Ministry smelt a rat and ordered JBC 'to discontinue all relations with any body except itself'—a circumlocution for requiring it to remove its Section D hat. JBC put up a stout resistance, but an investigation by Leslie Sheridan confirmed that MOI was being taken for a ride. In any case it seemed that most of JBC's output was cultural, and the rest was deemed to be of no great value. When he saw Sheridan's report CD enquired: 'How do we set about winding up this plaything?' It was decided to remove both JBC and United Correspondents from SO's payroll at the end of February 1941.

Section D spent the last months of peace examining 'the possibilities, limitations, and difficulties of covert propaganda' and continued to believe that in wartime SIS had a role to play. An organization would be needed to carry on propaganda activities that must not be attributed to the British government, for example, smuggling leaflets into Germany, and organizing postal propaganda to Germany from neutral countries. 'To reach the bulk of the population is best done by means of our organization.' To be effective propaganda must appear to be 'spontaneous and local. It must touch all sections of the population, including, may we emphasize, the local policeman in a small town perhaps 200 miles from the capital.'

Section D introduced a considerable quantity of its own propaganda into Germany through Rickman's organization. Fraser visited Sweden in August 1939 to study the problem. He found no enthusiasm for propaganda work among the dozen Swedes he met through his brother-in-law, and decided it would have to be handled by Britons. The local Reuters correspondent, who had been seven years in Sweden, would be ideal. His job took him all over the country and he was very willing to help. He knew a woman who ran a typing service and was prepared to duplicate propaganda material and to mail letters. Although some printers in Stockholm did stock German typefaces, Fraser felt none could be trusted with secret work. In Copenhagen he found a British employee at the airport, eager to co-operate, who offered to find other helpers, British and Danish.

Propaganda material was supplied from England by Section D and also produced locally—by Rickman, Brost, Biggs, and Behrisch. The last required an assurance that their objective was the destruction of the Nazis, not Germany. He had already been producing his own anti-Nazi propaganda and distributing it through Communist seamen. Berman-Fischer also co-operated. At the beginning of December 1939 London approved the employment of Ture Nerman, who could use his own printers. Unhappily Kurt Singer also came in at this time. Behrisch enlisted the help of the strongly anti-Nazi foreman of a printing works, who worked for him after hours.

Material was smuggled from Sweden into Germany in various ways. Behrisch recruited a commercial traveller under an obligation to him, who was on good terms with Danish Customs. He took to Copenhagen each week a batch of letters prepared by Brost. An agent there passed them on to a German skipper plying between Denmark and the German ports, who had long been engaged in smuggling. This man, who shared his profits with Customs officials, took 10,000 letters a week with fake German stamps, so they would seem to be posted within the Reich. Other distribution channels were explored but none was a going concern before Rickman's organization was blown.

These letters may or may not have affected German opinion, but at least they were successfully delivered. Behrisch addressed one to a friend in Germany who told him it had arrived safely. He also said some had fallen into the hands of the Gestapo who decided the enclosed leaflets could not have been printed in Germany since they were too professional for an undercover press. Section D thus learned the elementary lesson that leaflets purporting to be printed by a resistance group must look amateurish. One of Holdsworth's last signals to London reported that both Rickman and Biggs wanted to suspend the production of underground propaganda in Sweden. He and Tennant disagreed and had every intention of increasing the volume. 'I shall have to accept just what *XY* (Biggs) and *Ethel* (Rickman) can turn out because there aren't enough hours in the day for me to give much help to them.'

No less important than feeding propaganda secretly to Germany was keeping Swedish public opinion favourable to the Allies. At the end of 1939 Fraser asked Biggs to draw up a plan to influence Swedish newspaper editors, which could, of course, be attempted quite openly. The need for Section D to lend a hand on overt propaganda diminished when Peter Tennant was appointed Press Attaché, although it was believed there might still be occasions when his official position would make it difficult for him to do certain things. Then Rickman's group would be called on.

One approach was to attack the Swedish government for its 'pro-German neutrality' in the hope that public opinion would force a change of heart. Messages purporting to come from loyal Swedes were circulated, deprecating the government's attitude. For example: 'Neutrality? Nonsense! We are doing more than any other neutral country to prolong the world war and benefit the dictator states.' Again: 'On February 3rd the War Council of the allies agreed to send 50,000 men to save Finland. On February 16th Finland asked our government to allow this force through. On February 17th under pressure from German threats our leaders refused Finland's appeal. What have we gained by our cowardice? . . . An immediate future of danger and isolation.' And again 'it is not only Finland's fate that is being fought out amid her forests and lakes. If Finland falls, then Sweden's liberty, indeed all Scandinavian independence and the democratic way of life are threatened.' These messages were contained in leaflets and longer—perhaps too long—pamphlets, and invited the recipient to pass them on to 'your member of parliament, your doctor, your child's teacher, any influential public man you can think of, or just to a friend. It will all help. There is yet time to make good our betrayal of Finland.'

Anti-German propaganda for Swedish consumption at this stage concentrated on the occupation of Poland, the persecution of the church there, and the 'bestial brutality which characterizes Germany's rule'. 'If Germany wins we have seen that our independence will vanish . . . If the allies win, Sweden remains Swedish. No claim will be made on us except for our cooperation in building up a wiser and kinder Europe.'

Section D's only activity in Denmark was in the field of propaganda. Fraser contacted Ebbe Munck of the *Berlingske Tidende* in Copenhagen on 7 December 1939, when Munck agreed to provide information, although he did not want as yet to take an active part in any organization. He suggested possible lines of communication into Germany: a transport firm carrying fish there; a fleet of diesel-engined schooners based at Marstal, trading in the Baltic and occasionally going as far afield as England; crews of the Copenhagen–Malmö and Hälsingborg–Helsingør ferries who were expert smugglers, as were the crews and waiters on the Gedser–Warnemünde ferry. None of these suggestions was followed up.

Fraser visited Copenhagen on 7 April 1940 and was still there when the Germans arrived two days later. Being armed with a courier's pass, he was accepted as a member of the Legation and repatriated with their staff.

Guerrilla warfare

The first real opportunity for MI R, the guerrilla warfare specialists, to go into action came when the Germans landed in Norway in April 1940. Earlier, during the Russo-Finnish war (which broke out at the end of November 1939), MI R officers had played a part in helping the Finns. On 19 December 1939 Captains Andrew Croft and Malcolm Munthe were sent to superintend the off-loading at Bergen in Norway of aircraft and equipment sent by Britain for the use of the Finnish forces—a complicated task since many of the crates (ostensibly containing farm machinery, since they had to pass through neutral Norway and Sweden) were too big for the Bergen–Oslo railway and had to be modified. Later Croft and Munthe went to Tornio in Finland to hand over the equipment. They reported that the Finns were seriously short of aircraft. Three hundred more fighters would take the edge off the Russian air attack. The only good aircraft the Finns had were thirty French Moranes and thirty Gloster Gladiators. The fixed undercarriage of the latter—which made them out of date in Britain—was useful in Finland where it could easily be adapted to take skis.

Several other MI R officers went to Finland to act as instructors, including Second Lieutenants Whittington-Moe and Scott-Harston. They brought back information about the guerrilla tactics used by the Finns. Their white-uniformed ski patrols made them almost invisible against the snow and enabled them to operate well in front of their main defensive line, lying in wait for the Russian tanks. When the tanks were so near that they could not easily bring their guns to bear, the guerrillas would emerge from their snowy cover, or drop from the trees where they had been hiding, to throw Molotov cocktails and hand grenades into the rear ventilation hole of the tank. They also disabled tanks by pushing logs between their tracks and driving wheels. The patrols would cover as much as 200 miles before returning to base.

The British War Cabinet had endlessly debated the case for sending help to Finland. They got as far as preparing an expeditionary force, but the problem disappeared when the Finns capitulated on 12 March 1940. The troops were put on stand-by against the possibility that Germany would invade Norway to secure her iron ore supplies; but on

6 April an RAF reconnaissance flight spotted German warships heading for Norway's west coast and the chance of forestalling a German landing had vanished. The only hope now was that a British force could throw the Germans out before they consolidated their positions.

MI R, showing remarkable prescience, had been in no doubt that the Germans would invade Norway. It knew the COS had made no attempt to gather intelligence essential for a pre-emptive British invasion—reconnaissance of the ports, assessment of local reactions, study of lines of communication, and so on. Section D's study of coastal areas had not been built on (see pp. 45–6). Confident that sooner or later the War Cabinet must grasp the nettle and order a force to Norway, MI R took the only step open to it. A week before the German force sailed, it sent four officers to prepare the way for a British expedition: Torrance to Narvik, Croft to Bergen, Munthe to Stavanger and Palmer to Trondheim. The MI R War Diary records this as a pathetic attempt to fill the gap left by the War Cabinet and COS. Rather it was a brilliant planning stroke, unique at this sorry time. Had the other planners been equally far-sighted, the course of the war could have been very different.

As it was, these four young men found themselves facing the whole German invasion force. Captain Torrance, who had reached Narvik on 4 April, was awakened by the sound of gunfire on the ninth, which he assumed heralded the arrival of British troops; but the streets were full of German soldiers. After witnessing the engagement in which the Norwegian armoured ships *Eidsvold* and *Norge* were sunk with heavy loss of life, he made his way to a mountain hut. There he remained until 8 May when hunger forced him to return to Narvik. He was well looked after by the mayor and billeted in a private house until French troops arrived on 28 May. He then went north to Harstad on a fjord steamer to join the evacuation of the Allied force, which had failed to throw out the Germans.

At first Croft intended to remain in Bergen, where the Germans arrived in the early hours of 9 April, in the hope that an Allied force would arrive in due course. He walked to the top of the mountain railway looking in vain for British ships, then returned to the railway station where 'nine Germans were in possession . . . being admired by somewhat nervous spectators'. The British Consulate and his hotel were in the hands of the Germans so he abandoned his luggage and went to the shipping agents with whom he had worked in forwarding war material to the Finns. There he met Chaworth-Musters and another official from the Consulate, but they decided they would have a better chance of escape if they split up. Croft left on 10 April as soon as darkness fell. According to his account:

The new moon provided sufficient light, and dodging a couple of Germans, I got clear of the town. I had decided to make my way over the mountains to Samnanger Fjord, where I was told I could easily hide in some farmhouse until conditions were more stabilized. Actually, I still hoped to get on board a British ship either there or in Hardanger Fjord. Below 1,500 feet or so little snow lay on the mountains. The going was boggy but on the whole easy. I crossed over to Kallands Vand, then got over the main Os road and cut through the woods. Here I had an anxious 20 minutes standing motionless and being watched by a German picket. I had foolishly trodden on some dead twigs. Fortunately the man soon got bored and, the night being fairly dark, I was able to reach the narrow road which runs north-eastwards from Kallandseid to Harsgsdal. The 8 miles of road was entirely deserted and I now began to enjoy myself thoroughly.

Croft found the Germans already in control of Samnanger Fjord and had to make for Alesund. His journey was one of the more varied Odysseys of World War II. He covered 150 miles on foot, occasionally through waist-deep snow, and travelled at different times by car, lorry, motor boat, fjord steamer, and on borrowed skis. He was Norwegian to the Germans he encountered, and Swedish or English to the Norwegians. At Alesund he was taken on board the destroyer *Ashanti*.

Malcolm Munthe also had a remarkable journey. After helping the Consul's secretary (Miss Cragg, 'one of Britain's undauntable little heroines') who had stayed behind in Stavanger to burn the files, they joined the Consul, for whom the Germans were searching, and his family and staff, first in a farmhouse, and then in a more remote fisherman's cottage at the head of Fordefjord. Having waited in vain for a Norwegian officer supposed to contact him, Munthe left, hoping to find British troops. He fell in with a party of Norwegian soldiers and became involved in a skirmish with the Germans in which he was wounded in both legs and captured. He managed to escape some days later, and made for Bergen. At one point he found himself serenading a group of German soldiers with *Deutschland-uber-alles* rendered on an accordion in which he had hidden his papers. In Bergen he got a passport from the Swedish Consul in the name of Axel Axelson, Swedish by birth, a seaman in American ships all his life. Finally, many weeks later, he reached Stockholm and reported to the British Legation, where he was appointed Assistant Military Attaché in July 1940, and given the task of opening communication lines for the use of the SIS and the Norwegian resistance.

Palmer in Trondheim was less fortunate. He alone of the quartet was captured, being surprised asleep in bed by the Germans.

It was now accepted by the COS that MI R had an important part to play. On 12 April the Branch was instructed to reconnoitre Namsos to prepare for an Allied landing there. Next day a party of six led by Captain Peter Fleming flew to Namsos in a Sunderland flying boat which had to fight off two enemy aircraft. Their main role was to be to keep in touch by W/T with the approaching British expedition, but they found their sets were receiving only incoming signals. In any case, the Grand Hotel where they operated, and their W/T sets, were destroyed by enemy action on 20 April. They then gave some help on signals to Norwegian army headquarters, later said to have been invaluable. Their first message to a British ship (HMS *Calcutta*) was conveyed by manipulating the electric lights on a timber wharf, whereupon the ship provided them with signal lamps for more orthodox shore-to-ship contact.

Fleming's party remained in Namsos until the evacuation. They found their clothing and equipment deficient, lacking waterproof footwear, warm helmets, and snowshoes. They were also critical of the provision for the troops generally. The importance of white camouflage—linen and whitewash—had been overlooked. Given that warmer weather was on the way, men were handicapped by too heavy clothing. The two signals sergeants in the party considered their professional brethren 'quite inadequate'. Their own W/T sets were far too heavy. The hand generator was useless—tiring to operate and impossible to keep going for the whole of even a short message.

A second MI R party left Invergordon by Sunderland on 19 April—Major Jefferis of the technical branch, 'his wicked face lit with joy' (according to MI R's War Diary, which is more human than most) at the thought of the 1,000 lb. of explosives in his baggage; and Sergeant Tilsley. Their objective was to instruct the Norwegians in demolition work and to carry out demolitions themselves. After a day or two they were caught up in the action at Tretten on 23 April 'which was lost before it was begun, since the British territorial troops retreating up the valley had now for the most part been without both food and sleep for more than thirty-six hours, and had had no real rest for a week . . .'.[6] With the British position crumbling rapidly, the pair could no longer follow their trade and found themselves instead rallying parties of stragglers. They were finally evacuated by Sunderland on 28 April.

A third party—code-name Operation *Knife**—was supposed to go by submarine to the Sogne Fjord area to interfere with German communications. It left Rosyth on 23 April but the submarine (HMS *Truant*) was damaged by an enemy torpedo and returned to port. The operation was cancelled on 26 April.

The fourth and last party was dignified with the title No. 13 Military Mission, although it was but a three-man team: Major Brown, Captain R. B. Readhead, and their interpreter Corporal Dahl. They were told to report to the Norwegian Commander-in-Chief (General Ruge) and to encourage all aspects of guerrilla warfare pending the arrival of British troops. Brown must get in touch with the four 'Assistant Consuls' (i.e. Croft and his companions), who, with any other MI R officers still at liberty in Norway, would come under his command. Brown and Dahl reached Ruge's headquarters on 19 April, Readhead two days later.

The latter was ordered by Ruge to join a party of fifty ski troops operating from the mountains east of the Gudsbrandsdal to attack German communications. The general was opposed to demolitions, which he said would hinder his advance when British troops arrived, and would also lead to reprisals against civilians. In any case, the value of demolitions was doubtful since the Germans with unlimited timber at their disposal made short work of repairs. Brown had a similar task on the western side of the valley. Like Jefferis he became involved in the fighting at Tretten and was a focal point for British stragglers—eventually more than sixty of them. Although they had few weapons and were almost all without skis or snowshoes, they moved north parallel to the German advance and even inflicted damage on the enemy. When they heard on the radio on 2 May that Andalsnes had been evacuated by the Allies, they decided to split into small parties and make for the Swedish frontier. By common consent Readhead went ahead to facilitate their entry into Sweden. On the way—across 150 miles of marshes and mountain country with four major river valley crossings—he fell in with three other British soldiers whom he briefed to represent themselves as civilians. They hoped to cross the frontier unobserved, but were given away by a Norwegian they had met. To their surprise the Swedish police accepted they were civilians, and allowed them entry.

SOE's objectives in all the occupied countries through which the Allies might force their way back into Europe were threefold: to disseminate

* This operation is described by one of the members, Peter Kemp, in *No Colours No Crest* (Cassell, 1958).

subversive propaganda to weaken enemy morale; to create a patriot army to support a possible Allied invasion; and to establish sabotage groups to carry on a war of attrition against the occupying forces. It was some time, however, before these objectives, or at least the means of attaining them, were clearly seen.

A Section D paper of June 1940 on future strategy in Norway considered how 'to stimulate the civilian population to undertake guerrilla warfare against the German occupying troops, and ultimately to extend guerrilla tactics to open rebellion'. Preparations must be made for 'a major effort' in September and October 1940 before the really heavy snow became a serious obstacle for 'our troops'. The timetable was, of course, quite impossible. In August Section D was still looking forward to a major action in Norway. The best time now would be at the beginning of 1941 when the snow (which earlier would have hindered the patriots) would hinder the Germans. (Just why the snow would change sides was not made clear.) Support would come from coastal farming and fishing communities, and the professional and working classes in the towns. Fishing boats and submarines would take in arms concealed in herring boxes. 'When the ammunition, weapons and demolition materials have been smuggled in and the bands of Guerrillas are organized . . . some signal must be arranged so that all the people all over the country will rise at the same moment.'

The plan did not stop there. Flour, sugar, and coffee would be taken in to feed those working for the revolution, and pay for services rendered, since Norwegian currency would be hard to come by. 'Attacks by guerrillas during the dark, cold winter months would have a very strong effect on the already bad morale of the German garrisons', whose fighting quality was supposed to be impaired by the widespread use of drugs. If the supply was cut off, German resistance might collapse. 'To find one's billet fired in the middle of a dark, stormy night with deep snow all around is not a pleasant experience for the most hardened troops, and for young and drugged soldiers who are not used to snow conditions it would undoubtedly be disastrous.' This plan was still under consideration at the beginning of September 1940. Rebellion would be fomented in southern Norway where the population was greatest. Thirty to forty Norwegians would be recruited and trained in Britain and sent back to Norway to the area they knew best to build cells 'round which the general movement could grow'. The project would have to be discussed with the Norwegian government in Britain, not out of courtesy, but because it would be impossible to keep 'an operation of this magnitude' secret. 'The Norwegians as a people are notoriously ill-disciplined and are great talkers.' This wildly ambitious project was eventually written off, ostensibly because a current internal reorganization made it impossible to implement it.

It was left to Sir Frank Nelson to bring the planners down to earth. He recorded: 'In Norway we have at the moment literally no one in the field at all.' Six or eight Norwegians were undergoing training in Britain. Two others had been rejected for insubordination, and, since they must not be allowed to reveal details of the organization, they had been interned. 'I cannot see the possibility of any real activity either in sabotage or revolution in Norway for several months at least . . .' It is interesting that even Nelson contemplated 'revolution'. However, the idea of a countrywide rising gradually disappeared. It was accepted that only a full-scale Allied invasion could liberate the country.

Nelson's more sober assessment did not deter Dalton from continuing to paint a rosy picture of special operations in Scandinavia; and his wish to make the most of his organization's achievements communicated itself to some of his lieutenants, who also saw everything through rose-coloured spectacles.

All our reports go to show that, following the winter, Norwegian civilian morale in spite of, or perhaps because of, every kind of oppression and persecution by the Germans and the Norwegian quislings remains at the highest possible level. Our difficulty in fact (which is shared with the Royal Norwegian Government) is not to spur the people on, but to hold them back.

This over-painting of the picture was dictated by the 'political situation' within the organization, which meant the Minister's insistent demands for tangible results to lay at the Prime Minister's feet; but by the middle of 1941 the idea of immediate revolution in Norway had been finally put to rest. The policy now was to prepare 'for a simultaneous rising all over Norway on the occasion of, but on no account before, either a landing by the allies, or an incipient German collapse'.

This greater sense of realism was fostered by escaped Norwegians who gave reliable information about the state of affairs in the country and made possible a more accurate assessment of what could and could not be done. The most important was Captain Martin Linge, an actor by profession, but also an able officer who had been wounded during the German invasion. The Norwegian government appointed him liaison officer with SOE and he became the effective founder of the Norwegian Independent Company (NIC 1), or Linge Company as it was known informally. At first the members were civilians working under Linge, but in July 1941 were embodied in the Royal Norwegian Army. Between twenty and twenty-five men were recruited monthly until May 1943 when the total of 250 was deemed to be enough for the tasks envisaged. In the winter of 1944–5 fifty more were recruited to take part in the final operations in Norway. They were trained at SOE's first school (Station XVII at Brickenondbury in Hertfordshire), then at a Norwegian holding school at Fawley Court, Henley-on-Thames (STS 41) and finally, from November 1941, at Special Training School XXVI in the Aviemore district in Scotland between the River Spey and the Cairngorms in the heart of wild mountain country.

Denmark alone of Germany's victims in western Europe accepted the invaders without a struggle, and proclaimed her neutrality. For the first three years of the occupation the Danes were allowed to carry on unmolested. It was business as usual, or almost, for King Christian, parliament, the armed forces and police, so long as the agreement made with the Germans when they arrived was adhered to. In August 1941 the Danish government cemented the relationship by signing the anti-comintern pact. The Nazis hoped that, after a period of peaceful coexistence, the Danes would become willing supporters of their New Order and serve as an example to the rest of Europe, as they had hoped the Channel Islands would do. There was no Danish government in exile to plan resistance at home—nor was there any pressure on a peace-loving people to throw out the occupying forces. Patriotic Danes did set up a Free Danish Council in London in September 1940 but it was composed of members of the Danish colony whose names meant little in Denmark, and it could hardly be expected to rally the forces of resistance within the country. Many Danes overseas, including Count Reventlow, the Danish Minister in London, followed the King's lead. To take the law into their own hands might appear to undermine the monarchy, and call in question their loyalty. Not only had resistance to be built from nothing, the very idea of resistance had to be created.

At first Section D could do no more than encourage passive resistance in Denmark by means of subversive propaganda through radio, leaflets, and personal contact with Danes travelling abroad. There were some minor anti-Nazi activities within the country, but the people as a whole 'were not yet fuel that would spontaneously burst into flame'. They would have to be persuaded to hinder the German war effort, to help Allied raiding parties, to sabotage key installations; but they must never be made to think they were puppets dancing on an Allied string. They might take a long time to find their 'national consciousness', thanks to the benevolent attitude of the Germans; but sooner or later a national spirit must manifest itself. Those Danes who found their way to London to help the Allied cause, were, under the auspices of the Danish Council, drafted into the East Kent Regiment (The Buffs) of which King Christian was Colonel-in-Chief. This created a pool, akin to the Norwegian Linge Company, from which potential agents could be selected for training.

3
The Stockholm Mission: Sweden Section

CONTINGENCY plans for further special operations in Sweden were being made in April 1940, the month when Rickman and his associates were picked up by the Swedish police, and three months before Winston Churchill decreed that 'a new organization shall be established forthwith to co-ordinate all action, by way of subversion and sabotage, against the enemy overseas', to be known as the Special Operations Executive (SOE). Section D envisaged that a neutral Sweden would be used as a base to smuggle quantities of arms and personnel into Norway. If the country was invaded by the Nazis, or allied herself with them, British agents could no longer seek to direct activities there. Instead, Swedish trade unionists, most of whom opposed Nazism, and Social Democrat refugees from Germany would have to be recruited. Thirdly, if she threw in her lot with the Allies, defensive measures would obviously be taken by her own forces; but if the Germans gained a foothold in the country, groups recruited by Section D would operate behind German lines.

In September 1940 the War Cabinet, taking their cue from SOE's terms of reference, set out the principles on which special operations should be conducted world-wide. The stimulation of subversive tendencies in the occupied countries would contribute to the defeat of Germany by forcing Hitler to increase his armies of occupation and use up resources needed elsewhere. General uprisings coinciding with major Allied initiatives could be invaluable, but they must be held back until the time was ripe. A premature revolt would be counter-productive. All special operations must be kept in line with the overall strategic plan. A further paper showed how subversion could help strategy. 'Our aim in fact should be to get subversive activities laid on and ready for execution in all areas where there is any chance that they may be needed, so that wherever the fortunes of war may require action, the ground will be well prepared in advance.' The catalogue of areas where action might be required—Brittany, the Cherbourg peninsula, the south-west of France, Holland, Belgium, and south Norway—is interesting in the light of the eventual choice for D-Day. These papers[1] were merely guidelines, broad terms of reference, within which very little was at first accomplished.

Neutral Sweden posed special problems, especially in the aftermath of the Rickman débâcle. Although she was not an occupied country, she might be any day. Should, then, the fact of her neutrality simply be accepted? Should she be forced into the Allied camp, perhaps by the threat of sanctions, with the risk of throwing her instead into the arms of Hitler? Should SO 2 (as the sabotage component of the new organization had been christened, SO 1 being concerned with subversive propaganda) stir up an anti-Nazi movement within the country to compel the Germans to occupy it and further strain their resources? Section D had concluded after the German invasion of Denmark and Norway that 'the present feeling in Sweden may be summed up as a mixture of terror and shame, possibly coupled with a certain amount of gratification that the war has so far passed her by'. This was too great a simplification. The attitude of a nation cannot be defined so precisely. For example, members of the 'very big Fifth Column'—Section D's description of the pro-German element in the community—were certainly not ashamed of Hitler's successes.

In September 1940 Dalton instructed SOE's newly appointed head (CD), Sir Frank Nelson, to prepare a report on the organization. When he saw it his comment was 'Shocking!' Sir Robert Vansittart, at this time one of his advisers, agreed. There had been 'a pretty good mess!' In fact, in the summer of 1940 SOE had been too busy organizing itself, to do much more than evolve plans. Nelson's report said that in Sweden: 'We have nothing at all . . . There are at least two of our people in gaol; there is no field force, and no prospect of any . . . There was a D organization there early in the war, but owing to indiscretions of various kinds this has been dissolved.'

In reality, things were slightly different. On the debit side, there were four of 'our people' in gaol. On the credit side, there was a shadowy special operations presence. Peter Tennant, who had visited Stockholm in July 1939 'making certain enquiries' on behalf of the MOI, was in the autumn of that year appointed Press Attaché at the British Legation. Although he was not on Section D's payroll, his job kept him in touch with Swedish anti-Nazi organizations and he became in effect an honorary member of the Section. In July 1940 Malcolm Munthe arrived in Stockholm after his two-months' journey from Stavanger (see Chapter 2). He was given the cover of Assistant Military Attaché, and instructed to establish communication links with Norway to support the resistance there, and to provide intelligence for SIS. One organization Tennant kept in touch with, which had begun work early in the war, was entirely run by Swedes under the direction of Countess Amélie Posse, a leading anti-Nazi whom he first met in 1939. She managed a propaganda group centred on 'The Tuesday Club', which met weekly in Stockholm and had branches all over Sweden. A secret section concerned itself with W/T communication, paramilitary resistance, and sabotage—or at least the theory of these activities. Although The Tuesday Club had some financial help from SOE, it was at no time under its control. A second organization Tennant recruited himself. It consisted almost entirely of members of the Syndicalist Party, with objectives similar to those of The Tuesday Club, and was directly controlled by SOE. There was a chief organizer and a W/T expert in Stockholm, with subordinates located in the regions. Their job was to reconnoitre dropping zones, landing grounds, and targets for sabotage, and eventually to provide a countrywide W/T network. Their activities were purely preparatory, to make a blueprint which would be given substance if the Nazis invaded Sweden.

In October 1940 Charles Hambro, head of SOE's Scandinavian Branch, visited Stockholm to examine the position on the spot. It was decided that Tennant should become formally responsible for SOE work in Sweden, accounting to SO 1 for subversive propaganda, to SO 2 for subversive warfare, and to MOI for overt propaganda. Malcolm Munthe would continue to look after Norway. Ronald Turnbull, who had been Press Attaché in the British Legation in Copenhagen when the Germans arrived there, would go to Stockholm to be responsible for subversive operations in Denmark.

Dalton tried to meet the Prime Minister in January 1941 to convince him that he had tidied up the SO mess, and that the organization was now making good progress; but he 'found difficulty in compassing this'. That is to say, Churchill refused to see him. The Minister then prepared a long paper listing SO 2's achievements so far; but this also failed to penetrate the Prime Minister's defences. Sir Alexander Cadogan, Permanent Under-Secretary at the Foreign Office, did not pass it on, on the ground that 'it reads like a company prospectus . . . and I have a salt cellar by me when I study it'. The paper made mountains out of the few molehills SO 2 had so far turned up. It claimed, for example, that good progress was being made in the four Scandinavian countries (Finland being the fourth), in all of which agents had been planted, and good communications established—a very considerable overstatement.

In the spring of 1941 SOE headquarters, in consultation with the Foreign Office, studied a new approach in Sweden. The Swedish authorities would in effect be bribed by the offer of additional supplies of oil (their imports were controlled by the Allies) or up to 100 fighter aircraft, to accept SOE's presence in Sweden and to provide facilities for intelligence work. After lengthy discussion this exceedingly optimistic proposal was dropped. In May it was put to Mallet that, in view of the then likelihood that the Germans would attempt to occupy Sweden, the Swedes might be willing to co-operate in preparing to undermine the occupying forces. If resistance groups, supported by a network of W/T stations and arms dumps were organized now, they could do a great deal of damage to an army of occupation. Mallet refused to put the proposition to the Swedes, even informally; and London agreed to abandon it. Tennant suggested the Minister be told that SO 2 (the sabotage branch) was being withdrawn from Sweden, to be replaced by SO 1 (the more innocent subversive propaganda branch)—his ingenious idea being that SO 2 activities should carry on under the banner of SO 1. London rejected the ploy.

A month later the belief that a German invasion was imminent grew stronger; and it was feared that the Swedes, like the Danes and Norwegians before them, would fail to put up an effective resistance. It was felt that their will to withstand German pressure, never very great, was diminishing. They had just allowed the Nazis to send a whole division through Sweden to Finland, and to take over three Polish submarines lying in Swedish waters. It was, therefore, more important than ever to press on with plans for the post-occupational sabotage which the Swedes themselves seemed unwilling to contemplate. Because of the difficulty of external communication with Sweden—contact by sea was out of the question, and air drops would be too dangerous in the summer months with their long daylight—it was essential that there should be a first-rate W/T service both for internal use and to keep in close touch with headquarters in London. Four key stations were planned: at Gothenburg or Malmö in the south, at Stockholm, at Storlien close to the Norwegian border in central Sweden, and the fourth as near to the Finnish border as possible.

SOE's exasperation at the attitude of overseas missions is illustrated by a draft minute which Nelson prepared for Dalton to send to Clement Attlee, then Lord Privy Seal. Although the minute does not appear to have been sent, it is a most revealing document. Dalton's mind had been dwelling recently on the paramount necessity to speed up the infiltration of agents into countries which were still non-belligerent, including Sweden.

> What is required is a categorical direction from the Prime Minister or the Defence Committee, that whatever may be the inevitable objections raised by Ambassadors and Ministers . . . full, ungrudging and immediate support must be given to such Ministries and Organizations as may desire to send personnel for the purpose of organizing resistance to Axis penetration.

These people must be accepted without question, and reasonable requests for cover granted at once. Diplomatic staff must be ordered to co-operate in both the letter and the spirit of these instructions. If this could not be done, it would be useless to come to him (Dalton) 'when the conflagration is already under way, and expect my organizations to play any material part in the post-occupational ventures which would be of benefit to the allies'. This line, reprehensible in time of peace, but perfectly legitimate in total war against an enemy to whom no holds were barred, was echoed even by the Foreign Office. Cadogan, when dealing with the perennial squabbles between overseas missions and the secret organizations, wrote:

> The root trouble of all this is that we do not follow the much better (German) system of employing the Legation on subversive activity, SIS work and thuggery of all kinds. We have separate organizations working in absurd wasteful compartments which gives rise to all sorts of suspicion and misgiving—naturally—on both sides. I know it myself from experience abroad. I have a good mind to suggest that we should plainly and straightforwardly put our Heads of Mission in charge of all dirty work and definitely subordinate to them all these people with odd initials and numbers which puzzle me more than the enemy. (I am quite prepared to write a paper on the subject.)

Cadogan did not write his paper. British diplomatic missions never became all-embracing clandestine centres on the German model.

Mallet continued to be difficult. He opposed action by the Stockholm Mission in support of Operation *Claribel* (aimed at hindering a German invasion of Britain from the occupied countries) to organize the recruitment of guerrillas to attack German troops at the embarkation ports in Norway. Thereby he did his best to frustrate what might have proved a crucial operation. SOE appealed to the Foreign Office which criticized the Minister for his 'notorious over-anxiety about Swedish susceptibilities' and instructed him to co-operate. When Tennant arranged for someone to come from Britain to meet Swedish Trade

Unionists, potentially useful allies, Mallet complained that the visit had been arranged behind his back. In June Hambro said he was making a 'dead set' at SOE, and asked the Foreign Office to remind him that special operations were important, which they duly did. Malcolm Munthe's activities did not help. The operations planned in Sweden by his fictitious *Red Horse* embarrassed the Swedes, who knew what he was up to and feared German reactions. They made him *persona non grata* and he had to return to Britain, where he joined SOE headquarters. Once again Mallet was able to say 'I told you so!' (see Chapter 9).

In spite of the friction, SOE's Sweden policy remained unchanged. London reaffirmed that if the Germans did occupy the country Tennant must organize small groups to attack key installations of importance to the invaders. He replied that suitable men were to be found 'in intellectual and labour circles' and that he was already in touch with them. He had enlisted the co-operation of an organization secretly making W/T sets; and by March 1942 a W/T network was being formed throughout the country. The first successful link with London was established. An instructor was brought from Britain to speed up the work of training operators, but unhappily the specialist equipment he brought was found by Swedish Customs, and he was promptly returned to Britain. A Norwegian was recruited in his place to train the first batch of Swedish operators. In April Tennant visited London to discuss post-occupational work in Sweden; and was authorized to buy a yacht on his return to Stockholm. This was the *Valkyrie*, which was kept on Lake Malar near the capital. It was used to train W/T operators hidden from the eyes of the Swedish police, who continued to be highly suspicious of the British Legation and everything connected with it. *Valkyrie* was also used from time to time to dump compromising materials, and as a pleasure vessel by members of the Legation.

It was not until July 1942 that the COS provided a formal directive for SOE in Sweden. It confirmed that it should continue to build a clandestine organization to go into action if Sweden was occupied, or came under indirect German control. Stockholm drew up a plan to minimize the benefit the Germans would derive from exploitation of Swedish industry and services. Organizers with W/T sets able to keep in touch with Britain would be appointed in each of the ten districts (see map on p. 29) into which the country had been divided. Priority lists of sabotage targets were drawn up, with iron ore at the top. Then came merchant shipping and shipyards, railway communications between Sweden and Norway and between Sweden and Finland, train ferries from Germany and Denmark, engineering industry, and finally rayon manufacture. As it was thought unlikely that the Germans would attack before the spring of 1943, no more than this programme should be contemplated for the time being; and there must on no account be any attempt at pre-occupational demolitions of the sort expected from the Rickman organization.[2]

It was reaffirmed that activities must be purely exploratory and kept secret from the Swedish authorities. While pro-Nazi sentiment in Sweden would no doubt diminish as the Allied armies became more successful, and it would be easier then to enlist support for clandestine operations, it was believed that there were still powerful elements retarding the growth of pro-Allied feeling—including the Agrarian Party, some big landowners, bankers, industrialists, elements of the engineering and medical professions, small tradesmen and middle-class businessmen, an impressive catalogue. On the other hand there was no doubt that the anti-Nazi Trade Unions had many members prepared to join forces with SOE; and there was also The Tuesday Club, the Social Democrat Youth Movement, Fighting Democracy, and other groups with the same aims as SOE, although it was feared that, when it came to the crunch, they might contribute less than they had promised.

At this point the London Controlling Section (LCS) (the code-name for the War Cabinet's deception organization headed by Colonel John Bevan) offered SOE a deception plan with the objective of convincing the Germans that Sweden was about to join the Allies and forcing Hitler to occupy Sweden or greatly increase his occupation forces in Norway. He proposed building up a fake Anglo/American/Norwegian army in Scotland to convince the Germans that an Allied invasion was imminent, and that the Trade Unions should be encouraged to stir up anti-Nazi feeling within Sweden. Bevan, the principal exponent of deception in Britain, was delighted with his proposals. In submitting them he wrote: 'I shall be a wonderful stockbroker [his profession] after the war, and deceiving the entire City will be child's play—poor Hatry never knew the beginnings of it!' (Clarence Hatry was a businessman whose financial misdeeds became a *cause célèbre*.)

At the end of 1942 the organization in Sweden which the COS had approved was still very much in embryo. In October of that year Captain T. F. O'Reilly went to Stockholm at the request of Colonel Larden, then Assistant Military Attaché and head of the Stockholm Mission, to finalize plans for special operations in the event of a German occupation. Although it had been intended he should spend only three months on the job, he remained until August 1943, briefly replacing Larden as head of the Stockholm Mission. When he arrived, he was surprised to find that many Swedish newspapers and many of the Swedes he met were more confident of an Allied victory than was the man in the street in Britain; and this discovery coloured his attitude to special operations in Sweden. He believed the Swedes now had the intention and ability to offer formidable resistance to Germany, should it become necessary; and that the key to their continued neutrality was Allied success on the battle fronts, which would pay off better than anything special operations might accomplish.

It was, therefore, decided to concentrate on W/T matters rather than to try to build a full-scale sabotage organization, although several of the men with whom they were in touch were 'dynamitards by profession'. Major Croft, the other Assistant Military Attaché, had recruited several Norwegians qualified to give W/T instruction. Wilfred Latham held elementary courses in the use of codes and gave theoretical instruction in minor sabotage. There was some preliminary reconnaissance of dropping points. A proposal by Tennant that members of the organization should be issued with sabotage handbooks was rejected because some might be encouraged to use sabotage as a political weapon in time of peace. This risk would be run only when it was seen that German invasion was inevitable. The chief organizer was deemed to be a man of ability and drive, busy selecting additional district organizers. The principal W/T organizer was also competent, although for the moment his services were monopolized by the Swedish army.

A businesslike statement of assets was prepared:

the War Chest (funds provided for special operations in Sweden)

3 completed W/T

16 W/T under construction

Valkyrie, with outboard motor

one Dodge motor car

one chief organizer, plus sundry district organizers

one chief W/T organizer

two wireless constructors

5 Norwegian W/T instructors

20 W/T operators, 6 qualified, 14 training

It was admitted there was still no 'cut and dried' organization in any of the ten districts, and in three virtually nothing had been done. Although the chief organizer (who had been told 'in broad outline' the probable spheres of activity in the event of occupation) was travelling extensively to encourage progress, his movements were hampered by the interest of the police—a disquieting thought in the light of the earlier experiences of the Sweden Section. The country's increasing armed strength, and the greater opposition to Nazism which O'Reilly had sensed, made the need for an SOE presence more doubtful than ever. Moreover, in the judgement of the Ministry of Economic Warfare, even if Germany acquired all Sweden's resources and industry intact, it would not be catastrophic—a surprising departure from that Department's earlier view expressed to the War Cabinet, and from Winston Churchill's, that the interruption of iron ore supplies to Germany was of crucial importance to the Allies. The Foreign Office opposed anything more than exploratory work in Sweden. The loss of the country as a source of intelligence as a result of ill-advised special operations would be disastrous. Thus everything pointed to a soft-pedalling of SOE activities there.

A report in April 1943 claimed that the Swedish organization was still more of a skeleton than a body, and that it was making very slow progress. There was a good man in Malmö in District 1 who had a W/T and was training guerrillas. The organizer in District 2, centred on Gothenburg, had recruited a number of men, including several competent W/T operators, but he was too busy with his own work. In District 3 at Nassjo there was a small group which included a W/T operator. The position was much the same in the other seven districts. But in April most of what little flesh there was disappeared when the Swedish police arrested the chief organizer, the chief W/T organizer, and several other members of the organization, without implicating any of the Legation staff, although the connection must have been well known. According to O'Reilly: 'As a result of the police surveillance of our organization and of Latham and myself we were able to compile descriptions and a list of the registration numbers of no less than 15 Swedish secret police cars'—for which information SIS was most grateful. In June 1943 it was finally accepted that the threat of a German invasion need no longer be taken seriously; and the next directive to the Sweden Section virtually ended special operations in Sweden—as distinct from operations into Norway and Denmark based on Sweden. The directive also led to a clash between CD (head of SOE), and C (head of SIS).

The former was told that in no circumstances must he extend his existing communications network in Sweden; and in the same breath that he must provide trained W/T operators with sets able to transmit to Britain. He complained that one paragraph ordered him to find W/T operators, and another ordered him not to do so. When this was put to C, who had an extensive communications network in Sweden, he replied curtly:

> The crux of the matter so far as I can see is: does it advance the war effort to risk our and your valuable communications through Sweden to existing agents in order to make preparations to meet a German invasion of Sweden, a contingency which the Joint Planners recently declared to be remote? Personally, I feel quite sure it is not; and in this connection it will be remembered that post-occupational schemes have never so far operated when the time came, and that agents who have promised so much in the comparative security of peace have not fulfilled their promises in war.

This was rather below the belt, although it cannot be denied that, if the Sweden Section had suddenly been put to the test by a snap invasion of Sweden, it would have made little contribution to stemming the tide, or even harassing the Germans after they had established themselves. The most that SOE was now allowed to do *vis-à-vis* Sweden was to earmark possible new W/T operators on the understanding that no approach was made to any individual—a totally meaningless concession. In November 1943 Sporborg said he saw no reason for retraining even a dormant organization in Sweden, and, in spite of opposition from Wilson, in charge of Sweden at headquarters, who argued that a Swedish group should be retained to report on road conditions, commercial undertakings helping the Germans, and other barrel-scraping activities, special operations within Sweden came to an end. From now on the SOE staff in Stockholm concentrated on supporting activities in Norway and Denmark, for which Sweden provided an admirable base, and in helping in the series of blockade-running expeditions described in the next chapter.

It is convenient here to deal with SOE's minor interest in two other countries in the Scandinavian region: Finland and Estonia. It was originally contemplated that there would be full-scale activities in the former—it has been mentioned above (see p. 61) that early in 1941 Dalton implied that SOE was already a going concern there. In fact it was not until April 1941 that Captain Alfred Merry (formerly a commercial diplomat) arrived to take up the duties of organizer. The intention was to appoint twelve'trade inspectors' ostensibly to supervise imports authorized by the British Ministry of Economic Warfare, who would be located near the German officials working under the Transit Agreement with Sweden. Merry found men for all twelve posts and he himself took the title of 'Controller of Trade Inspectors'. It was mooted in July that two W/T operators should be sent from Sweden, but before any progress could be made the arrival of the Germans led to the closing of the British Legation in Helsinki. Merry, his deputy Alfred Cochrane, and all the trade inspectors were withdrawn. It was hopefully suggested they should make arrangements for some continuing SOE effort in Finland, but it proved too difficult. Almost the only action taken by Merry in the few weeks he was in the country, apart from an investigation of the Petsamo nickel trade, was negative. The Stockholm Mission had suggested that a Finnish ship might be used to block the iron ore harbour at Lulea in Sweden, and Merry was asked to explore the possibilities. He pointed out that it would be necessary to enlist the support of the captain of the ship and the owners, which would certainly not be forthcoming. A further complication was that the ship was part of a government-controlled pool. So this solitary proposal vanished without trace. Although SOE headquarters claimed that 'work in Finland was to be by no means abandoned', in the event it was; and Merry was told to destroy his records on 12 June 1941, probably in anticipation of the German attack on Russia.

SOE fared no better in Estonia, where there was a single operation. Ronald Seth, who had been teaching in Estonia before the war, proposed a sabotage attack on five shale oil mines along the Tallinn–Narva railway. His estimate of the men—and money—required for his mission was worked out with commendable detail.

	Pounds
To the wrecking of 5 oil-shale mines:	
Regular payments to special organization of 40 men at Kr. 40 per month for 8 months	640
To watching and probable action against the port of Tallinn	
Cell of 5 men at Kr. 40 per month for 8 months	80
(Identical provision for the ports of Paldiski and Parnu)	160

	Pounds
To watching and probable action against the railway junction of Tapa	
Cell of 5 men at Kr. 40 per month for 8 months	80
To organization of cells of 3 men at various points along the Tallinn–Riga and Tallinn–Narva–Leningrad railway lines and probable action against	100
To watching and probable action against the airfields at Viljandi and Ratvere	
2 cells of 3 men at Kr. 40 per month for 8 months	160
To my own subsistence for one year at Kr. 80 per month	50
To reserve for contingencies which may arise from time to time	230
Total	£1,500

Currency: £1,000 in dollars in 5 dollar notes, numbers not to run consecutively. £450 in Swedish kronen, numbers not to run consecutively. £50 in gold Russian roubles and German mark pieces, if possible.

Seth was dropped in a clearing in the woods on the Leesi promontory on 24 October 1942. According to the dispatcher's report, he was 'quite happy and jumped without hesitation and well'. The three containers he took with him all landed safely and he flashed the 'all's well' signal on his torch. In a report dated 2 September 1944 Seth provides an account of his adventures which leaves the reader in no doubt that fact can be stranger than fiction, as the following short extract shows:

I have no time now to give even a brief account of my activities, except to say that I was betrayed in Estonia, after having lost all my equipment. I have been tortured by the Gestapo, condemned to death, actually had the rope round my neck, but the trap refused to work, reprieved, sentenced to death a second time, but managed to gain the confidence of two Gestapo men in Frankfurt-on-Main who passed me to friends in the Luftwaffe Intelligence Department who planned to send me to England as a Nazi agent . . . Before I was captured in Estonia, by a miraculous stroke of luck I got hold of 18 of my altimeter switches [devices to detonate a charge when an aircraft reached a certain height]. With these I blew up 3 Fiesler-Storch aircraft, and an anti-naval gun with 13 soldiers, 4 large 5,000-litre [1,100-gallon] oil tankers (motors) . . . In Paris I shot 2 soldiers in the Metro on 1 January 1944 . . . I was tortured in order to be made to give the numbers of my radio crystals, but managed after 17 hours to out-tire my torturers.

Seth's remarkable mission, although it failed in its main objective, did show a remarkable profit; and it also showed that there are exceptions to the generally accepted rule that no mortal could withstand a full-scale Gestapo interrogation.

4

Sea operations: Sweden

AFTER the occupation of Denmark and Norway the Germans blockaded the Skagerrak to cut off all Swedish ports from the North Sea. Britain could no longer import goods under the Anglo-Swedish War Trade Agreement, in particular badly needed special steels—a double blow since the Germans would be delighted to take them over. Section D and George Binney, British Iron and Steel Control representative in Scandinavia, thought the blockade could be run when the nights were long; and between January 1941 and the winter of 1944–5 there were five operations to do this: *Rubble*, *Performance*, *Cabaret*, *Bridford*, and *Moonshine*.

Operation *Rubble*

Section D was encouraged by the voyage of a small Swedish vessel (the *Nora*) to Britain in May 1940 in spite of the blockade. Binney was authorized to charter a Finnish ship (the *Lahti*) to carry 300 tons of steel for the aircraft industry—but she was spotted by a German plane and meekly allowed herself to be shepherded into Kristiansand. Undaunted, Charles Hambro, Section Head Scandinavia, and Harry Sporborg arranged for Binney to buy large tonnages of the materials most urgently needed, ostensibly to be stored in Norwegian vessels lying in Swedish waters, but in reality to be taken through the blockade when conditions were favourable. The Norwegian captains of the selected ships were unwilling to co-operate—hardly surprising as all were on full pay with nothing to do but enjoy themselves. One was positively hostile. He sent an *en clair* telegram to Oslo telling his owners what was in the wind, which inevitably found its way to the German police.

Another problem was how to man the ships. There were in Sweden the crews of several British iron ore carriers caught at Narvik, and survivors of HMS *Hunter* and *Hardy* sunk there—all free to return to Britain if they could find the means. After a good deal of persuasion enough volunteers came forward—mainly British, with about one-third

Norwegians and a handful of Swedes. Of the five captains, three were British and two Norwegian.[1]

The Chief of the Harbour Police in Gothenburg, Captain Ivar Blücker, was well-disposed towards the Allies. He was instrumental in stopping the Germans from planting Nazi sympathizers in the crews; and for the arrest of a Swedish naval officer who informed them about the ships' movements. When the fleet finally sailed, Blücker arranged for local telephone lines to be cut so that the fact would not be immediately reported to the Germans. Nevertheless, it was obvious to their agents what was going on.

The loaded ships were assembled in Brofjord, north of the small port of Lysekil, whither they were escorted by Swedish warships to discourage attacks by German patrols violating Swedish territorial waters. Here they were painted battleship grey, and lay at anchor, for a time in solid ice, awaiting a favourable weather forecast—which meant as much fog as possible. On 23 January 1941 Blücker said the ships should get under way immediately; and Binney agreed. The vessels under his command were the *Elizabeth Bakke* (5,400 tons, Captain Andrew Henry); the *John Bakke* (4,700 tons, Captain W. J. Escudier); his flagship, *Tai Shan* (6,900 tons, captained by the Norwegian Einar Isachsen); *Taurus* (4,700 tons, also with a Norwegian captain, Carl Jensen); and finally the tanker *Ranja* (12,000 tons, Captain J. Nicolson).

The voyage is best described in Binney's words:

It was snowing in the Brofjord. The *Ranja* went out first at 3 p.m.; she was the slowest ship; moreover she was in ballast, and if a ship was to be torpedoed at the outset it was better to lose her salt water ballast than to lose a steel cargo. Then came the *John Bakke*, followed closely by *Taurus* and the *Elizabeth Bakke* . . . Finally we left about 4.30 p.m., dropping our pilot and Captain Blücker off Lysekil at 5.45 p.m. It was dusk and still snowing slightly with visibility of about one and a half miles . . . No one slept that night . . . all eyes were straining through the darkness for some half-expected attack . . . At 8 a.m. it was light, and the snowclad Norwegian coastline, of which we were 28 miles south, was clearly in view. Every moment we expected the Coastal Command patrol, but we later learned that owing to bad weather conditions at home it had been unable to start . . . We still had approximately 55 miles to go before reaching the rendezvous with the naval escort and we anticipated serious trouble from German air patrols. It was Mediterranean weather with blue sky and calm seas. Our salvation lay in the fact that no air patrols were active.

Shortly after noon the promised Royal Navy escort hove in sight. German submarines and aircraft came on the scene at last but were driven off with only one casualty, the Swedish First Officer on the *Ranja* being fatally wounded when the ship was machine-gunned from the air. At 6.30 a.m. on 25 January the fleet reached Kirkwall in the Orkney Islands.

It was not only the snow in Norway which grounded enemy aircraft for several hours that saved *Rubble*. In November 1940 agents had reported to Berlin that there would be an attempted break-out; and the German naval staff were desperately keen to intercept the ships. They believed it would be difficult due to uncertain weather conditions, pack ice, the long nights, and the relatively high speed of the Norwegian ships. At first their naval patrols were committed to routine tasks that had priority over a constant watch on Swedish waters, like a cat waiting for a mouse to emerge; but on Christmas Day, when they thought the break-out was imminent, patrol boats were put on stand-by watch. After 11 January, however, the bulk of German naval forces in the area were required to concentrate on getting the battleships *Scharnhorst* and *Gneisenau* safely through the Kattegat and Skagerrak on their way to the North Sea and Atlantic. They left Kiel on 22 January and two days later reached the North Sea, by which time Binney's fleet was west of Stavanger. When it was eventually spotted it was too late to do much about it.

SOE suffered the not unusual fate of a secret organization: much of the credit for *Rubble* went elsewhere. When the Deputy Prime Minister (Clement Attlee) said there were differences of opinion as to who had played the major part, Dalton pointed out that no one had mentioned the contribution of his organization 'but this is to our credit, since it shows that we are pretty good at covering our tracks!' The truth of the matter, he went on, was that Charles Hambro had conceived the plan and organized it, ably assisted by Harry Sporborg, his second in command. Dalton added that, of course, some credit was due to others, including Binney. If the award of honours is a reliable indicator, Hambro made the greater contribution. After much Ministerial debate, Hambro became KBE, Binney a plain knight. Blücker was rewarded with a pair of cuff links bearing the royal crest.*

Binney himself had no doubt who was the star. In a curious account written in the third person, widely distributed to the dismay of his

* Not the pair the King was wearing when he dubbed Hambro, as Ralph Barker suggests in *The Blockade Busters* (p. 75) but a pair taken from stock and sent through the Foreign Office by the King's Private Secretary (FO 371 29410 (1.4.41)).

colleagues, in the view of the paramount need for secrecy, he recorded:

> The protagonist in this adventure, well-known in these Islands, and in many countries overseas as a businessman of considerable enterprise and ability, and not giving the impression at first sight of being a twentieth-century buccaneer, fair-haired, blue-eyed and of medium height, he is not the type that enjoys publicity . . . [2]

Which is presumably why he attributed his eulogy to a hypothetical third person, who continues: ' "Are you planning to give an encore?" I asked. "Hope springs eternal," was Binney's reply. "But don't forget that, though you use the same pack, the cards fall differently every deal." ' Alas, the truth of this was to be proved by the succeeding operations.[3]

Operation *Performance*

There remained in Swedish waters other Norwegian vessels, including five tankers, four merchantmen, and a factory whaling ship. Encouraged by *Rubble*'s success, Dalton asked for a special flight to take Binney back to Sweden to prepare *Rubble II*; but the Norwegian government in Britain objected, ostensibly because the nights were getting too short, the Germans would be more vigilant, and there was no cargo immediately available.[4] In reality they were afraid of precipitating a showdown between the Swedes and the Germans which might have unpleasant consequences for Norway.[5] However, after discussion, Trygve Lie, the Norwegian Foreign Minister, approved the operation so long as Hysing Olsen, Chairman of the Norwegian Trade and Shipping Mission, thought it feasible. Olsen said it was, given air and naval protection. There was a strong case for making the attempt, since the *Rubble* material would be used up before the end of 1942. Binney was told to go to Sweden right away to buy 15,000 tons of special steels.[6]

Since the operation could not be carried out until the autumn when the nights would be longer, Binney had plenty of time to make his purchases and to arrange for captains to come from Britain—six of them. Others, and the crews, were found in Sweden as they had been for *Rubble*. The Germans, furious at *Rubble*'s success, threatened that if one more Norwegian ship escaped to Britain no Swedish vessel would pass their blockade. They obtained by duress power of attorney from the original owners of the Norwegian ships (which had been requisitioned by the Norwegian government in London) and started legal proceedings to take them over, helped by a deliberate change in the Swedish law.

Although their case failed, they contrived by this manœuvre to keep Binney's ships tied up in port for several months.

Churchill was personally interested, and chose a code-name for the new operation, one of his favourite pastimes. His *Oudenarde* confused everybody concerned as they were still thinking in terms of *Rubble II*, and finally a fresh start was made with *Performance*.[7] The Swedes, under great pressure from the German government, now urged Binney to call off the operation on the ground that it could not possibly succeed. Ribbentrop summoned the Swedish Minister to Berlin to say he was astonished by the line taken by the Swedish hostile press. The Führer could not understand it at a time when Germany was fighting for Europe's future. Some Swedes might like to sit comfortably in the chimney corner simply observing the European situation, or flirting with Britons who shared their sentiments; but all the nations of Europe, even France, were behind Germany in her attack on the disease-spreading cesspool of the Bolshevik menace. Only Switzerland and Sweden were holding back. It was incredible that Sweden should stand idly by while the fate of Europe for a thousand years was being decided. The Führer was deeply wounded by her attitude. Having delivered the gospel according to Hitler, Ribbentrop said that if the Norwegian ships were allowed to leave it would be seen as a most hostile act. Finland had never put legal quibbles before her vital interests.[8]

When Bjorn Prytz, the Swedish Minister in London, poured out all this to Eden, the Foreign Secretary suggested that the German threat should not be taken seriously. Prytz agreed, but admitted his government took a different view. The Germans claimed that Sweden was favouring Britain and that the Norwegian ships would 'be juggled into British hands'. They now demanded they should be returned to their Norwegian owners without legal procedure.[9] In Stockholm, Erik Boheman, Secretary General of the Swedish Foreign Ministry, made another effort to halt *Performance*. He agreed that the Swedish navy should escort the ships within Swedish waters, as they had done for *Rubble*, and that, if the Germans attacked them, Swedish forces would take the appropriate action. However, the Norwegian ships would not be allowed to remain stationary. They must run the gauntlet of any German warships lying in wait, or return to Gothenburg, mission unaccomplished. For good measure he said he had no doubt the steps the Germans were taking would stop the ships from getting through.[10]

It was not until March 1942—a year after Trygve Lie had approved the operation—that *Performance* got under way. The difficulties to be overcome, even before it contended with ice, minefields, *Luftwaffe*, and *Kriegsmarine*, were formidable. Binney complained bitterly that the Swedish authorities had striven to keep the ships in port 'by fair means or foul' until the nights would be too short. They 'methodically deprived us of every element of surprise for our escape. As they well knew, and freely admitted, we were a prey to German espionage . . . within 15 minutes of our weighing anchor the news would be in Oslo, Copenhagen, and Berlin.' He held a briefing meeting on his flagship, *Dicto* (a 5,000-ton merchant vessel, Captain D. J. Nicholas) in the afternoon of 31 March 1942. Except for *Dicto*, the ships carried both a British and Norwegian captain, it being intended that the Norwegian should take over in international waters. The masters of the five tankers were: Captains Calvert and Blindheim on *BP Newton* (10,300 tons); Gilling and Monsen (*Rigmor*, 6,300 tons); Small and Reksten (*Buccaneer*, 6,200 tons); Reeve and Bull-Neilsen (*Storsten*, 5,300 tons); and Nicol and Trovik (*Lind*, 460 tons). The other three merchantmen, in addition to *Dicto*, were commanded by Captains Kershaw and Schnitler (*Lionel*, 5,600 tons); Nicolson and Seeberg (*Gudvang*, 1,500 tons); and Donald and Nordby (*Charente*, 1,300 tons). Finally, the 12,300-ton factory whaling ship *Skyttern* had Captains Wilson and Kristiansen. In the interests of security, only the British captains were invited to the briefing meeting, perhaps an error of political judgement on Binney's part.

It was unanimously agreed to sail that evening keeping within territorial waters until the fog that was forecast gave them cover in the open sea. Everything was in order: deck hoses to deal with any fires; Lewis guns smuggled into Gothenburg had been mounted; charges to sink the ships to avoid capture; drugs to keep the W/T operators awake for the whole voyage. Binney reminded the captains they were undertaking an evasive operation, not the charge of the Light Brigade.

He signalled London that they were about to leave so that the RAF and Royal Navy would be ready to receive them. It then transpired that the fog forecast was suspect; and he ruled that, if the weather did prove unsuitable, the whole fleet should return to Gothenburg. Finally he issued an order of the day concluding:

> So let us Merchant Seamen—400 strong—shape a westerly course in good heart, counting it an excellent privilege that we have been chosen by Providence to man these ships in the immortal cause of freedom. God speed our ships upon this venture. Long live King George! Long live King Haakon!

The convoy safely reached the open sea where coastal ice forced the

Dicto out of territorial waters. She was immediately fired on by a German armed trawler, and sought sanctuary near Hallo Island to await a later and hopefully better weather forecast. When it was still unfavourable Binney signalled the fact to the other ships with the implication, in the light of his earlier ruling, that they should return to port, but none received the message. *Dicto* then made her way back to Gothenburg. The second ship, *Lionel*, was refused permission to lie in Swedish waters (as Boheman had warned) by the ships escorting her. She, too, was attacked by a German patrol and forced to return to Gothenburg.

The patchy fog meant that every now and then the other ships became sitting targets. *Storsten* was delayed by faulty steering gear, but put to sea on the morning of 1 April. She was almost clear of the Skagerrak, 30 miles south of Kristiansand, when she was attacked by a German bomber and patrol boat, and torpedoed amidships. She was abandoned after her scuttling charges had been blown. The crew and the refugees she carried set a course for Scotland, Reeve and thirty-one others in a lifeboat, Bull-Neilsen and eighteen in the motor boat. They were spotted next day by an RAF plane; but before they could be rescued Reeve's boat was driven ashore at Jossingfjord in Norway, where they were betrayed to the Germans by quislings. Eight managed to escape to Sweden. The motor boat was never heard of again.*

Four other ships driven out of territorial waters became easy targets and had to be scuttled: the *Charente*, *Gudvang*, *Buccaneer*, and *Skyttern*. The tanker *Rigmor* made a gallant attempt to reach Scotland. The early morning of 1 April found her sandwiched between a Swedish destroyer trying to make her leave Swedish waters, and German patrol boats hoping to sink her as soon as she did. About 10 a.m. the weather closed in and it seemed safe to make a run for it. By evening she was in the middle of the Skagerrak east of Kristiansand when she was hit by a bomb from a German aircraft. Gilling was wounded by a machine gun bullet. Nevertheless *Rigmor* continued to make reasonable progress and when next morning she was sighted by RAF planes the crew assumed their troubles were over. But after the planes left two more German aircraft attacked. A bomb ripped open her starboard side and a second put one of her engines out of action. As she now listed heavily and was unnavigable, the order was given to abandon ship, whereupon two more Nazi planes appeared and *Rigmor* was hit amidships by a torpedo. The enemy aircraft were driven off by four British destroyers which picked up *Rigmor*'s crew from the boats. An attempt to tow the stricken ship was

* For its possible fate see Barker, p. 134.

abandoned because of the heavy swell, and she was sunk by gunfire. SOE saw it as a great misfortune that the RAF was unable to protect its prize. It lost a fine ship and, although she carried only 600 tons, it was material for the manufacture of aero engines, desperately needed by the RAF.

Only two ships reached Scotland. The tiny coastal tanker *Lind* evaded patrol boats and aircraft in the Skagerrak. On 2 April she was attacked by a plane which was driven off by the RAF. A second German aircraft sent two torpedoes harmlessly beneath her. Thereafter air cover and a destroyer escort enabled her to reach Methil Roads in Fifeshire on 4 April. She had only a few hundredweights of ball-bearings—her main purpose had been to act as a 'lifeboat' for the other ships—but her cargo was worth its weight in gold.

The other ship which managed to get to Britain was the *BP Newton*, the fastest and second largest of Binney's fleet with 5,000 tons. In the early morning of 1 April she was ordered to put to sea by two Swedish destroyers, but ignored the order until it started to snow and it seemed safe to make for the open sea. As she did so she met *Lionel* returning to Gothenburg, and was warned by her of the dangers ahead, but her captains decided to press on. About 1 p.m. she was fired on by a small armed trawler, but no damage was done, and her superior speed soon took her out of range. Shortly afterwards two more armed trawlers attacked, but were equally unsuccessful. Later in the afternoon she was attacked by three German aircraft, which did no damage, being kept at a safe distance by *Newton*'s Lewis guns, which shot down one. Early on 2 April she was picked up by a destroyer and an RAF plane and reached Leith without further incident.

This time the Germans had been able to concentrate a greater force against the blockade runners. Their agents had hinted at a break-out as early as July 1941, but the naval staff discounted any attempt in the summer. In November, when the Norwegian ships were still detained pending the outcome of the court case, they relaxed their patrols; and when the ships were free to leave in the middle of March 1942 (so far as the Swedish courts were concerned) there was very little German strength in the Skagerrak. Had the blockade runners been allowed to leave immediately after the court decision they might well have got clean away; but, according to the German records, by the end of March 'the blockade line was held in depth, which contributed to the success achieved and justified the long patient vigil maintained by the patrol vessels through a period of 9 months.' The German naval staff attributed the delay in leaving to ice conditions, apparently unaware that the Swedish authorities were playing their game for them. The only small crumb of comfort from the British point of view was the time taken to report the break-out. A German patrol boat had spotted the convoy at 3 a.m. on 1 April but failed to report the sighting until 8 a.m. An immediate report 'would have enabled counter-measures to be taken much sooner and perhaps every one of the Norwegian ships would have been accounted for'. The Germans grumbled that their Naval Attaché in Stockholm had not been told about the break-out until 11 a.m. on 1 April 'which showed a deliberate lack of prompt news reporting by the Swedes'.

The fact that the *Performance* ships had been armed was widely publicized by the press in Britain and by the BBC, leading the Germans to accuse the Swedes of complicity. Boheman summoned Mallet and Binney to tell them that this latest exploit was most embarrassing. Binney accepted responsibility and said afterwards that, in their meeting with Boheman, Mallet 'gave me precisely the support which the situation required, by restrained gesture of surprise, pain and despair at my confession of guilt'. The Swedish Cabinet was worried by the German reaction, and the fact that Britain's fortunes were at a low ebb did not help. In recent months she had lost the *Prince of Wales* and *Repulse* in the Far East; Hong Kong, Singapore, and Rangoon had fallen; Rommel was carrying all before him in North Africa; the vast operation to sink the *Gneisenau*, *Scharnhorst*, and *Prinz Eugen* as they made their dash through the Channel from Brest to Germany had failed dismally at a staggering cost of forty-two aircraft. Given this catalogue of disaster, it was natural that Sweden should seek to refute any suggestion that she was acting against the interests of the all-conquering Axis; and to placate the Germans the Swedish authorities encouraged a press campaign 'against our gun running, not only dragging my name [Binney's] but that of the Legation through the mud'. The Norwegian government in Britain was also upset. With the benefit of hindsight they argued that the Norwegian captains had not been fully consulted, and that Royal Navy and RAF support had been negligible.

Operation *Cabaret*

The importance of maintaining supplies of ball-bearings led Binney in July 1942 to plan a third operation to run the German blockade, code-named *Cabaret*. He believed that the relative failure of *Performance* must have reduced enemy vigilance, and that *Lionel* and *Dicto* could be safely picked up by two destroyers outside the Skagerrak. He accepted they would be attacked by armed trawlers, torpedo boats, E-boats and from the air. Further, a channel would have to be swept through the minefields. Defensive armament must be sent with the destroyers. The Naval Staff wrote the plan off as hopeless. The convoy would be at the mercy of shore-based aircraft, quite apart from the surface vessels. SOE persisted something must be done and it was finally agreed to attempt *Cabaret* in December 1942, on the understanding that two motor gunboats (MGB)—less vulnerable and more expendable than destroyers—would be employed as escorts. They were 110 feet long, relatively heavily armed, with a draught of only 9 feet.

Ten thousand tons of materials were bought and by the end of November *Lionel* and *Dicto* were fully laden. Base plates were bolted on their decks so that Oerlikon and Lewis guns brought by the MGBs could be instantly mounted. Paravanes were fitted to give protection from mines. The bridges were shielded with sandbags. Towing gear was installed to rescue the MGBs if they ran out of fuel. CD advised against the use of conventional demolition charges. Instead 'some excellent little gadgets' produced by Station XII which could easily be carried on the MGBs 'would ensure a good and safe hole'. Special W/T code messages (*Stratum 1–6*) were devised to let the Swedish organization know what progress the MGBs were making across the North Sea. Agents in Kristiansand were asked to procure the latest information about minefields.

While the operation was being planned in London by Hambro and Binney, and in Sweden by Waring in the Legation, and Coleridge in the Consulate at Gothenburg, the Foreign Office strove to get the Swedes to facilitate the ships' departure. Although the Swedish authorities claimed to be strictly neutral, they still tended to favour the more successful belligerents. Mallet wrote:

> We had got into a pretty good position here until this disaster in Egypt [Rommel's advance to El Alamein] but now I can foresee a cooling-off among all but the real stalwarts . . . If only we could occasionally win a battle we might find the Swedes more ready to be cheeky to Adolf.

In November Churchill told Roosevelt that the Swedes were refusing to clear the two ships whose cargoes were badly needed. It might do the trick if Boheman, then in Washington negotiating Sweden's oil quota, were told that any increase would be conditional on the ships' release.

The President agreed the oil quota should be doubled, clearance of

Lionel and *Dicto* being one of the conditions, but when the State Department put this to Boheman he claimed he had made it clear in London that the ships would not be released, and implied that the British had accepted it. In fact, he had suggested in London that someone should go to Stockholm to discuss the position. Later he admitted to the Americans that Sweden had no right to hold the ships, but added that equally there was no obligation to give export licences for their cargoes. The longer he stalled the better from Sweden's point of view, since *Cabaret* would become increasingly dangerous as the nights became shorter, and might have to be called off.

SOE asked that Sweden be given an ultimatum. The State Department agreed, provided it was deferred till Boheman (travelling by Clipper flying boat, which could take many days) got back to Stockholm, and also until two oil tankers bound for Sweden had arrived. Their reasoning was that if *Lionel* and *Dicto* were released the Germans might close the Skagerrak in reprisal before the Sweden-bound tankers got through. The United States would then be accused of trickery in granting the oil quota and then creating conditions which made its delivery impossible. SOE accepted this, although it meant further deferring *Cabaret*.

In Stockholm the Ministry of Foreign Affairs confirmed to the British Legation that there was no question of letting the ships go. It was no longer a matter of law—arming the *Performance* ships had made it a political question. Later Boheman claimed the real trouble was that the Prime Minister deeply resented the arming of the ships as he had personally assured the Germans they were unarmed. A friend of the Prime Minister, who had supplied some of the ball-bearings, said the Swedish Cabinet refused to believe the cargoes were really important, and that Britain would never take drastic action over them. On the other hand, they were fully convinced that, if the ships were allowed to escape, the Germans would close the Skagerrak.

On 16 December Mallet paved the way for the ultimatum by formally asking permission for the ships to sail. The Swedish officials were 'surprised and pained' by the statement that Boheman (now marooned in Bermuda) had been warned in Washington of the serious consequences if the Swedes did not co-operate. He had told them nothing of this. Boheman was taking so long to get back that it was agreed with American blessing to deliver the ultimatum on 21 December, before the Swedish Cabinet dispersed for Christmas. It recited that the Royal Swedish Government admitted there was no legal obstacle to the departure of the ships; but that Britain would not request clearance until the two oil tankers had arrived safely. If the ships were not released then, the British and United States governments would consider themselves free to withdraw from their agreement about import quotas for Sweden.

The Minister of Foreign Affairs, forced to face the moment of truth, complained that the Allies were being most unfair. Sweden had to trade with Germany in order to exist. Now she was being asked to do something that would bring her into direct conflict with the Nazis. He held out no hope that Sweden would comply with the Allies' demand. When this was reported to Eden, he said he was 'not much impressed by the lamentations and prognostications' that the Germans would cut off Sweden's imports. They still needed her iron ore. Mallet did not agree. The Swedes genuinely feared a trade war with the Germans, who had good stocks of iron ore; and they were very conscious that Hitler could flatten large areas of Stockholm and Gothenburg, as he had done in Rotterdam. Nevertheless Mallet was instructed to tell the Minister of Foreign Affairs that Britain had every right to cut off Sweden's imports as a reprisal for her denial of Britain's rights. Since the Swedes boasted that they had never bowed to a German threat, it followed that their numerous and serious breaches of their neutrality obligations had been committed, not out of fear, but out of benevolence towards Germany.

On 11 January the Swedish government faced up to neutrality. Boheman (who had at last found his way back from Bermuda) told Mallet that *Lionel* and *Dicto* were free to leave, but only on certain conditions. The SOE teams in Sweden were overjoyed, and then deeply shocked when London said 'No conditions!' CD answered the protests by explaining that so long as the Swedes were genuinely afraid of the Germans it was possible to respect their viewpoint; but 'when they with true Swedish materialism proceed to spoil their fine action by attaching conditions designed to take advantage of our need, and force us to give them material concessions, they forfeited all claim to our consideration'. He added that everyone concerned in London deprecated the Swedish attitude. The reply came:

> We have never felt more disappointed. Unexpectedly and at the last moment cargoes are within our grasp if we choose. Agreement to Swedish conditions could not by any stretch of the imagination damage our interests . . . Of course they are not entitled to conditions but by the time the inevitable discussions are concluded it will be too late.

On 15 January the Swedes finally gave in. *Lionel* and *Dicto* could leave, but not before 17 January, when the second of the two oil tankers was due in Swedish waters. Dangerously late in the season, *Cabaret* was in a position to start. The ships had been fully prepared since early December. An agent in Norway had supplied essential information about recent changes in coastal lights, and in the shipping regulations off Bergen and Kristiansand. The harbour authorities had been very co-operative. Not so the naval authorities. They refused permission for compass adjustments to be carried out in advance; and, since they insisted the ships should be moored inside the *Drottningholm*, which would have to be moved before they made their final departure, there was no chance of slipping away unobtrusively. In the meantime the crew of the *Drottningholm* had to cross *Lionel* and *Dicto* when they went ashore, and could see everything going on on the two ships.

The Swedish authorities still played the neutrality game by giving the Germans advance warning of the movements of the ships. In any case, German agents reported they had gone to Hakefjord to prepare for their escape, although they had moved at dead of night and in dense fog. Armed German trawlers began to patrol the coast for the first time since July 1942, and four destroyers put into Kristiansand. Binney, realizing that the odds were against success, proposed a deception plan. When the ships left Swedish waters they would report that *Lionel*'s engines had broken down, and that *Dicto* was towing her back. The Legation would instruct them—*en clair* for the benefit of the Germans—to return to Gothenburg. In fact, the ships would make for the open sea at full speed, hoping to be well on their way before the Germans tumbled to what was happening. Nothing came of this. A proposal to create a diversion by making a false start was also ruled out. It was believed that the German destroyers waiting at Kristiansand would shortly move away; and a false start would certainly keep them there.

A conference in London on 30 January decided that, so long as the German destroyers did remain at Kristiansand, there would be no question of launching *Cabaret*. Then on 3 February agents reported that the destroyers had moved. The two MGBs (commanded by Lieutenant-Commander H. W. Duff-Still, RNVR, and Lieutenant P. Williams, RNVR) would set out next night (4 February). There was tremendous rejoicing among the crews of *Lionel* and *Dicto* whose morale had steadily declined during the interminable delays—until they received the signal *Stratum 6*: 'MGBs returning to base owing to unfavourable conditions.' They had got to within 70 miles of the mouth of the Skagerrak, when they were recalled, partly because of bad weather, but mainly because at the time a German naval force, which *Cabaret* would have run straight into, was passing through the Skagerrak. London instructed *Cabaret* to remain on twenty-four hours' notice. On 9 February Waring complained that they had been told nothing more and that 'this state of affairs is profoundly depressing'. Ever since 17 January the weather had been perfect for the operation, but now the days were 1 hour 40 minutes longer and the moon was waxing.

On 24 February Binney decided *Cabaret* was no longer feasible. Those concerned in Sweden felt badly let down; and thought it might have been worth while for the two ships, or at least the faster of the pair, to make a dash for it unescorted. The only consolation London could offer was the Admiralty's estimate that the *Cabaret* threat had forced the Germans to deploy as many as twenty-four ships to prevent a break-out—and that had been a valuable service.

Operation *Bridford*

By the end of February 1943, almost before he had written off *Cabaret*, Binney was planning a new attempt—Operation *Bridford*. His idea was that the MGBs assigned to escort *Lionel* and *Dicto* should themselves act as cargo vessels. Stripped of internal fittings they could carry 40 tons, not much compared with the 16,000 waiting on board the Norwegian ships, but better than nothing. They could operate with two officers and sixteen men (instead of their normal complement of forty), who would have to be carefully selected because of the mental and physical strain imposed by heavy weather in the North Sea. Their size made them difficult to detect on radar. They could ward off attacks by E-boats. Destroyers might be a danger, but the MGBs with their shallow draught could play hide and seek with them in minefields. In ten hours of darkness they could cover 200 miles which would take them beyond the range of German shore-based aircraft in Norway. Because of their cargo of steel they would need gyroscopic compasses—an advantage since the time taken to adjust magnetic compasses made it easier for the waiting Germans to intercept them. Their main problem was that they were designed for short coastal voyages and it was uncertain whether they would stand up to round trips of 1,000 miles through the North Sea in winter. The crews would be provided with SOE's special self-heating soup to avoid cooking, and the standard RAF pack of Horlicks, chewing gum, and benzedrine.

It had to be decided whether they would sail as warships or as coastal craft. The White Ensign (making them warships) gave certain privileges in neutral waters, but there were balancing restrictions. The Red Ensign (making them merchantmen) entitled them to the privileges accorded to all merchant ships: access to Swedish ports; the right to anchor at night round the coast; and the right to carry defensive armament. It was at first considered that the balance of advantage lay with the White Ensign, until Boheman belatedly pointed out that a recent royal decree banned all belligerent ships from Swedish waters unless they were in distress. So the five MGBs selected for the operation became Ellermans Wilson coastal craft with merchant navy crews. In his unbounded enthusiasm Binney saw the beginnings of a regular merchant service taking in ports from Stromsund on the Norwegian frontier in the south to Lulea in the Baltic in the far north.

Mallet was instructed to prepare for the operation. He must leave the Swedes in no doubt that, if they had cleared *Lionel* and *Dicto* in November 1942, as they were legally bound to do, before the German destroyer patrol had been laid on, the ships would easily have got away. Further, they had put *Cabaret* at risk by refusing permission to adjust compasses in advance, and by warning the Germans of their impending departure. However, Britain recognized the problems set by the ban on shipping in and out of Gothenburg (a German reprisal for the Swedes' supposed complicity in *Cabaret*) and suggested that they should tell the Germans they had now compelled Britain to lay up the Norwegian ships permanently. The chances were that the Germans would agree to the resumption of the Gothenburg traffic and withdraw their destroyers for service elsewhere, leaving the coast clear for the MGBs. If this happened, Britain would expect the Swedish authorities to facilitate the new operation.

As usual Mallet was a prophet of doom. The Swedes would not help. Or they would insist on increased import quotas, which would mean negotiations with the United States 'with the usual consequences of delay and haggling'. He preferred a conciliatory approach. The failure of *Cabaret* was simply due to bad weather and the vigilance of the Germans, not to the unhelpful Swedes. Only if Boheman proved recalcitrant should a big stick be waved. In the event the Swedes fell in with the British proposals more readily than Mallet expected. It was by now by no means certain that the Germans were going to win the war.

One of the British requirements was that *Lionel* and *Dicto*, from which small parcels of ball-bearings were to be transferred, should move from Gothenburg to Lysekil, the port to be used by the MGBs. It had become a restricted area after the *Rubble* affair, which made it more difficult for

German agents to follow what went on there. In spite of objections from the Swedish naval authorities, *Lionel* and *Dicto* were allowed to anchor in the nearby Brofjord.

After intensive crew training during the summer, the five MGBs left the Humber on 26 October with Binney on the *Nonsuch* (Captain W. H. Jackson) as Commodore. The others were *Gay Viking* (Captain H. Whitfield); *Gay Corsair* (Captain R. Tanton); *Hopewell* (Captain D. Stokes); and *Master Standfast* (Captain B. Goodman). Next morning *Gay Viking* lost touch because of engine trouble. During the day the other four were twice unsuccessfully attacked by German aircraft, and that evening Binney decided, partly because of the difficulty they were having with their engines, that they should all return to England. Meanwhile *Gay Viking*, in spite of her engine trouble, was making her way to Lysekil, where she arrived to a rapturous welcome from a group including the Mayor and the local customs officer.

Perhaps moved by *Gay Viking*'s success and the belief of some of his captains that they too should have pressed on, Binney prepared to leave for Lysekil on 31 October, the day *Gay Viking* got back. As soon as they sailed, *Gay Corsair* developed engine trouble and had to return to port. *Hopewell*, with Binney on board, and *Master Standfast* with G. R. W. Holdsworth as captain in place of Goodman, who had been injured on the earlier voyage, reached the Skagerrak without incident. The pair lost touch and at 4.00 a.m. *Master Standfast* was picked up by a German patrol boat which opened fire and fatally wounded the captain. The Germans boarded and towed the MGB to Frederikshavn. This encounter was a piece of bad luck since the Germans had only two patrol boats operating in the Skagerrak at any one time, one reason why *Gay Viking* had been able to come and go unmolested. At the end of October the number was increased to six and *Master Standfast* had the misfortune to run into one of the new arrivals.

Six other successful round trips were made. Three by *Gay Corsair* (arrived Lysekil 6 December 1943, 21 January 1944, and 8 March 1944); two by *Gay Viking* (28 December 1943 and 9 March 1944); one by *Hopewell* (December 1943); and one by *Nonsuch* (17 February 1944). Although the Ministry of Supply had hoped to receive 400 tons of special materials, only 280 tons were brought to England. It was decided to call off *Bridford* in the middle of March, although by this time it might have been possible to bring out *Lionel* and *Dicto* unmolested. The Germans had abandoned patrols against the blockade runners because all the available craft were needed for other tasks; and at the end of

March even the *Luftwaffe* had stopped their Skagerrak reconnaissances. The German Naval Staff accepted that, if the Norwegian ships broke out, they could not be stopped.

In spite of the objections of the naval authorities, the MGBs were allowed to lie alongside *Lionel* and *Dicto* in Brofjord which simplified the transfer of cargo. The crews were welcomed ashore, invited to film shows, given hot baths, and generous meals without coupons. Everything was done to keep them hidden from German agents.

> Without exception our presence was genuinely welcomed by the civilian population: the pilots, the fishermen, the naval ratings, the Customs, the police and harbour officials, and all the simple folk who have had so little voice in the wartime government of Sweden.

The naval authorities continued to be as difficult as they dared. They demanded three hours' notice of the departure of the coasters, which would obviously help the Germans, but were beaten down to one. At one point a shore battery opened fire on one of the MGBs, which drew apologies from individual naval officers but not from the high command. On another occasion the navy sent 'post office engineers' on board *Hopewell*, ostensibly to explain what parts of the W/T should be sealed while she was in Swedish waters, but actually to examine the radar equipment on behalf of the *Kriegsmarine*, one of the 'post office engineers' being head of the Swedish subsidiary of the German electrical concern AEG. The Prime Minister and Boheman both apologized for this incident and promised that disciplinary action would be taken.

Operation *Moonshine*

When *Moonshine*, the last of the blockade-running operations was mounted, there was virtually no blockade to run. The purpose was to carry to Sweden for onward carriage to Denmark arms urgently required by the Danish resistance movement; and to bring back from Sweden the sort of materials the earlier operations had brought. The plan was discussed by Binney, Lieutenant-Commander Hollingworth (head of SOE's Danish Section), Ebbe Munck, the Danish resistance leader who had been closely associated with SOE's team in Stockholm, and Lieutenant-Commander S. B. J. Reynolds, in the first week of August 1944. The Danes had a vessel, the *Mariana*, plying between Sweden and Denmark ostensibly to bring refugees from Denmark but actually smuggling Swedish sub-machine-guns to Denmark with the blessing of 'certain high personages' in Stockholm, further evidence that more Swedes had now decided the Germans could not win the war. Two elaborate cover plans were evolved to hoodwink the Swedish authorities.

Under the first, Munck would tell his Swedish friends that the Germans planned to massacre a large number of Danish patriots before they finally pulled out of Denmark. There was no point in bringing these people out right away, since the Nazis would have no difficulty in finding substitutes. So the resistance had got hold of a ship lying in a Jutland harbour which would bring them all out on the eve of the massacre. This ship needed spare parts for its engine; and these would be smuggled to the *Mariana* from the *Bridford* MGBs. Alternatively, Munck would make a direct approach to Gustav Moeller, the Swedish Minister who was stage-managing the export of machine guns to Denmark with the connivance of the police and Customs officials in Hälsingborg. He would point out that the machine guns were useless without ammunition, which Britain would be happy to supply in innocent packages brought in by the MGBs. This was deemed to be the more convincing ploy—but all that the Swedish authorities were to be told was that the MGBs were there to collect more ball-bearings.

Nonsuch and *Hopewell*, whose engines had been modified during the summer, were loaded with 26 tons of arms destined for Denmark, and looked like getting through at the first attempt when they were recalled because enemy ships were reported in the Skagerrak. *Hopewell* came back with a broken crankshaft, and, since the engines were still unreliable, *Gay Viking* was brought in in the hope that two out of the three might get through. Eventually, after no fewer than twenty unsuccessful outward voyages, all three reached Lysekil on 15 January 1945 with 44 tons of arms for Denmark—over 1,000 carbines, nearly 1,000 Sten guns, smaller quantities of Bren guns and bazookas, and 2 million rounds of ammunition—all boxed as paint, steel wire, assorted bolts, fireclay, and so on.

On the return trip in February, *Nonsuch* carried 33 tons of ball-bearings. Unfortunately, in thick weather in the Skagerrak *Hopewell* rammed *Gay Viking*, which had to be scuttled. After repairs, *Hopewell* got safely back to Britain with 30 tons more of ball-bearings. Both MGBs made further efforts to get arms to the Danes but were frustrated because of weather or engine trouble; and in March 1945 *Moonshine* was finally called off.

Reynolds, who commanded *Moonshine* as Binney was sick (and who went under the name of Brian Bingham, because of his association with the earlier operations), was struck by the changed attitude of the Swedish naval authorities, who now went out of their way to be helpful. The MGBs were allowed to go to Gothenburg without dismounting their guns. There were cheers from the workmen all through the docks, and waiting on board *Dicto* was a deputation consisting of the Governor of the County, the Head of Customs, and the Head of the Secret Police. The Chief of Naval Staff 'asked whether we had any complaints, and was everything possible being done to help us? This was an unheard of honour when one reflects upon their previous attitude!' Reynolds was summoned by the Admiral, West Coast, who wanted to facilitate their departure, which 'must be arranged between us personally'. A far cry from the days when the Swedish naval staff 'appeared to place every obstacle in our path to try to sabotage our efforts'. Nothing succeeds like success!

5

Sea Operations: Norway

IT might be thought that the mountains of Norway would afford ideal bases for guerrilla activities on the lines successfully followed in the Balkans. The Hardangervidda and Juttenheim regions in southern Norway, for example, contain vast tracts that could never be cleared of guerrillas without a totally disproportionate use of manpower. But the terrain has little in common with the mountains of Greece or Yugoslavia, where guerrilla groups could live in comparative security. Parties did survive in Norway in the mountains under German occupation, but it was impossible to build large bands able to threaten the enemy. Postal and telephone communications were easily monitored, and couriers had to run the gauntlet of patrols on the few roads, or keep to mountain tracks where movement was slow. The rugged terrain made parachute landings dangerous, and new arrivals were soon spotted in the sparsely populated valleys. Moreover, air supply, hazardous at all times, was ruled out from May to September because of long daylight; and at first there were few aircraft available. These conditions made the coastal areas more promising ground for SOE's initial activities, although the North Sea is anything but an easy supply route at any season of the year; and it was the coastal areas that were developed first. By the end of the war, however, large numbers of W/T stations and guerrilla groups were being maintained all over the country, in the coastal areas mainly by boat, and, thanks to the great expansion of the Allied air forces, further inland from the air.

The Shetland Bus Service

Most of the sea operations were to land agents and supplies, to lay the foundations for a clandestine army, and to bring out agents and refugees, including resistance men wanted by the Gestapo. They were almost all mounted from Shetland, the point in Britain nearest Norway, to which in 1940 large numbers of Norwegian fishing boats brought men anxious to carry on the fight against the Nazis (see Appendix 1). It seemed to SO 2 and SIS that these men could make a useful contribution in their particular fields by providing a ferry service to and from Norway; and, after a joint investigation on the spot, Major Mitchell of SIS was appointed to set up and manage the service. Four recently arrived fishing boats were chartered, volunteer crews recruited from the refugee fishermen, and a large farmhouse—Flemington, 3 miles from Cat Firth, where many of the fishing boats lay—was acquired to house Mitchell and his agents, and to act as a training school. For the time being the crews continued to live on board their boats. Flemington was chosen because of its lonely situation in a remote valley, equidistant from Lerwick (the only town), Lunna Voe, and Scalloway, so that training could be carried out in complete privacy. Since the crew members were civilians, whose services were voluntary, they were paid £4 a week, with free food and clothing, plus a bonus of £10 for every trip to Norway. They were free to decide whether or not to go on any particular trip, and also to sever their connection with the Base at a week's notice.

Management of the Base was not easy. Life in Shetland in the days before North Sea oil was as rugged as the Islands themselves, and weather and wartime conditions did not help. Shetland is about 200 miles from Aberdeen, the nearest mainland port able to provide the stores needed by the Base and its agents; but the movement of the ordinary supply ships was erratic. They could not be relied on to bring the specialized items just when they were needed. Therefore, a drifter was acquired from the navy to ply between Aberdeen and Scalloway exclusively on the Base's business. This service continued until 1944 when the drifter was withdrawn because of the prior claims of D-Day. One of the larger fishing boats was then pressed into service and did the job very well. Agents were sent to Shetland by air to spare them a trying sea voyage immediately before embarking on an even more trying voyage at the worst time of the year; and, because of the very limited accommodation, their movements through the Base were spread as evenly as possible.

The first operation left on 22 December 1940 when the fishing boat *Vita* (Skipper Ingvald Johansen), manned by the crew that had brought her to Shetland, went to Langoy in the Bergen area to land an agent. He was safely put on shore, but was then spotted by the Germans and had to return to the *Vita*, which luckily had not left, and which brought him back to Shetland. There were twelve more trips, mostly sucessful, before the end of May 1941 when the service had to be suspended because of the almost continuous daylight. (Of the 203 operations carried out during the whole war, none was attempted in June or July, only nine in August, and eleven in May.) Fifteen agents were landed, and eighteen men of the resistance and thirty-nine refugees brought out. The summer of 1941 was devoted to preparing for greatly increased activity in the next winter season. There were now twelve boats, mostly in need of repairs, which were carried out by a Lerwick firm. The increased traffic made it necessary to move to a better anchorage at Lunna Voe, where a large house was acquired to house the crews.

The first trip of the new season sailed on 30 August 1941, and eleven more were made in September. Four of them re-established contacts in the Bueland area to the south of Alesund, and left quantities of arms there. Agents were also infiltrated into the Alesund area and for the first time agents were brought out—two by the *Vita*, which also brought eight refugees. Later the *Igland* (Skipper Ole Grotle) had a narrow escape when the party to be contacted at the head of Vigefjord—too far from the open sea for comfort—was arrested before the *Igland* arrived, and she had to beat a hasty retreat. Shortly afterwards the Base suffered its first loss, *Vita* being captured with her crew at Reko as a result of German penetration of the organization assembling refugees. The refugees escaped, but the Germans were lying in wait for *Vita*, and captured her crew, who spent the rest of the war as prisoners. When news of this misfortune reached the Base, *Aksel* (Skipper Bard Grotle) was already on her way to the same area. Aircraft were sent to turn her back, and an improvised warning was included in the BBC news, but she went on unaware of the danger. Luck was with her, however, for, after spending two days at Traenan, she was able to return to Shetland unmolested.

During this period more operations were undertaken than the Base could readily handle; and at the end of September only one boat—*Siglaos*—was wholly prepared for sea. The local engineers, unfamiliar with Norwegian engines, had found it difficult to cope with the frequent breakdowns. A further problem was that routine work, even the cleaning and maintenance of the quarters at Lunna Voe, was done by the crew themselves in the interests of security. They very reasonably resented having to 'do' for themselves in this way, perhaps while they were preparing for a dangerous and arduous voyage, or having just returned from one. Hambro and Sporborg visited the Base to see what could be done and as a result of their visit the establishment was increased and it was arranged that the Admiralty would become responsible for maintenance of the boats. Before these changes could improve matters, the Base was plunged into a particularly severe winter; and between October and December 1941 only ten trips were possible.

In October *Siglaos* (Skipper Peter Salen) successfully landed an agent for Haugesund. *Nordsjoen* (Skipper Leif Larsen) laid forty-two mines in Edoyfjord but was damaged by heavy weather on the return journey and had to be abandoned near Kristiansund. The crew got ashore and, after lying low for three weeks near Alesund, stole the fishing boat *Arthur* and brought her safely back to Shetland. *Siglaos* later landed an SIS agent in Bomelo but was attacked by aircraft on the way back and suffered a fatal casualty and serious damage which imposed a further strain on the repair facilities. At the end of October there were thirty-three boats in varying stages of disrepair; and, in spite of the agreement with the Admiralty, the local naval resources could not help much. A hurricane blew for six days from 11 November doing enormous damage to many of the boats. Some of those laid up were driven ashore, several becoming total wrecks, and others requiring many months of repair work. *Arthur*, returning from a successful trip to Bueland, was hove to for five days, during which she lost one man overboard. *Blia* (Skipper Leroy), returning from the Haugesund area with a large party of refugees, was lost with all forty-two souls on board. After January 1942 there was some improvement. The repair and maintenance squads began to catch up with the backlog and from January to May there were twenty-two sailings, virtually all successful.

Major Mitchell was recalled to London and his place was taken by Major Arthur Sclater.

> Both men were outstanding, Mitchell for his work as a pioneer, Sclater for his firm handling of a machine which gradually became more complex both in itself and in its relations with outside authorities. A word of tribute must also go to Mrs Sclater, whose personality at Flemington made it something more than an agents' hotel, and did much to keep the morale of parties leaving for the field at high level.

In the summer a start was made with putting the organization on a more formal footing. Better accommodation was provided at Scalloway, to which all personnel moved, except for some administrative staff who remained in Lerwick. The slow rate of boat repair had been largely due to the fact that there was only a single slipway in Lerwick which might be occupied for weeks on end by naval craft. In November 1941 the Admiralty agreed to provide a slipway for SOE's exclusive use, but nine months passed without anything happening. SOE therefore got permission to do the work itself, and completed the new slipway in less than three months, ready for use in October 1942; but the absence of this facility in the summer months meant that once again repairs were still well behind hand. This meant that only two trips were sailed in September and two more in October, in addition to Operation *Title* (see pp. 102 ff.). There were eight in November when the weather was better than usual, all successful. There were high hopes that everything would now be plain sailing.

Aksel (Skipper Bard Grotle), returning from the Smolen area, sent the code signal 'Send help' from a position 200 miles north of Shetland. Aircraft and a motor torpedo boat (MTB) were immediately sent to her help and next day a Catalina spotted the ship sinking, and her crew in dinghies. Unhappily the MTB could not get to them as she had to return to port for fuel, and when other surface craft reached her reported position neither they, nor two of the Base's own boats, nor more aircraft found anything. The search was abandoned after four days. The cause of the *Aksel*'s loss was never discovered. The weather had not been unduly bad, which left the possibility that she had struck a floating mine, or been attacked by enemy aircraft. It was a serious loss, for Grotle was a veteran of the Base, 'a most forceful and popular character', and his crew were very experienced. *Sandoy*, which had sailed two days after *Aksel* (on 8 December 1942) to the Traenan area, was attacked by aircraft and presumed sunk with all hands. Early in 1943 *Feie* bound for the Bergen area from Burghead was lost without trace. Her skipper was Ole Grotle, brother of Bard.

In spite of these setbacks there was no shortage of volunteers for future trips; but the success rate was far from satisfactory. Of the seventeen operations carried out between January 1943 and the end of the season in March, nine were unsuccessful, for a variety of reasons: for example, spotted by enemy aircraft, picked up by coastal searchlights, heavy weather making a landing impossible. In one case a reconnaissance party was put on shore on the wrong island and walked straight into a German outpost. Most serious of all was the loss of two more vessels, the *Bergholm* (Skipper Leif Larsen) and *Brattholm* (Skipper Sverre Kverhellen). The former, which had sailed from Shetland on 17 March bound for the Traenan area, was sunk after a two-hour engagement with enemy aircraft. The skipper and crew rowed to Alesund where they made contact with the UK through the *Antrum* W/T station. One had died, and all except two were wounded. Their compatriots in Shetland wanted to go to their rescue, but they were persuaded that it would make better sense for a force of much faster MTBs to do this job, which they did successfully three weeks later. The *Brattholm*'s crew had a less happy ending. Their vessel was bottled up in a fjord by a German patrol boat, and their attempt to blow it up when the enemy came alongside unhappily failed. In the ensuing engagement most of the crew were killed. Two who survived were later shot as spies. The only man to escape was the agent the *Brattholm* was delivering, who although wounded reached Sweden after suffering incredible privations.

It was now apparent that fishing boats were no longer viable on the 'Shetland Bus routes'. The Germans had gradually improved their coast-watching and defences in the regions where the Shetland boats operated; and their hand was further strengthened by the fact that most of the large boats in the Norwegian fishing fleet were laid up through lack of oil, which made it easier to spot visitors from Britain. If the Base was to continue operating it must have faster and more heavily armed boats; and SOE spent the summer of 1943 trying to find them. Coastal Forces' craft, fishery cruisers, and other alternatives either could not be spared or lacked the required speed, range, or seaworthiness; and for a time it looked as if the Base would have to close, with disastrous consequences for those agents already established in Norway. Hambro again came to the rescue. He persuaded the United States Navy to transfer three submarine chasers to the Base. They were ideal for clandestine operations: much the same size as MTBs but able to weather the worst the North Sea could throw at them. They were shipped as deck cargo to the Clyde where their American crews trained men from the Shetland base before the vessels were handed over at the end of October, ready for the 1943–4 season.

The whole outlook now changed. Although it was possible to fit in only two trips before the end of 1943, the final total for the whole season was thirty-four, of which twenty-six pinpointed their target exactly on time—infinitely better than the fishing boats were capable of. One consequence of the introduction of the sub-chasers was that the bonus hitherto paid to the crews was deemed no longer appropriate, which led to considerable dissatisfaction. The men were still civilians, although in 1942 the Norwegian high command had appointed a naval officer to look after their welfare and discipline, and had agreed that they should wear Norwegian naval uniform as members of the Norwegian Independent Naval Unit—the counterpart of the Norwegian Independent Company (NIC 1). SOE warned the men that they would be foolish to make a fuss about the loss of the bonus, on the ground that if they withdrew their voluntary service they would find themselves in the Royal Norwegian Navy (RNN). They persisted in their claim with the inevitable result. Although nine left the Base, the others were recruited by the RNN, and received naval rates of pay. One advantage was that SOE now had access to Norwegian naval personnel generally, not merely volunteer fishermen, and it became easier to man their vessels.

The final season—the winter of 1944–5—was by far the most successful. Out of eighty trips, seventy-six achieved their objective. The main reasons were the undoubted suitability of the sub-chasers and improved communication with the resistance groups in Norway; but important contributory factors were the decrease in German air activity and surface patrols. It was now possible to penetrate far into the fjords with virtually no opposition; and, in addition to its primary job of servicing the resistance movement, the Shetland Base craft continued to bring out large numbers of refugees. The records show how much the service had improved. In the two seasons 1941–2 and 1942–3, when the average number of fishing boats available at any one time was six, the Base carried out eighty operations, landed sixty-two agents, and 165 tons of stores, and picked up seventy-two agents and refugees, at a cost of eight boats and nearly fifty lives. In an equal period—the next two winter seasons—the three sub-chasers carried out 114 operations, landed 128 agents and 220 tons of stores, and picked up 271 agents and refugees—with no loss of life, or craft.

Mission Antrum

The Shetland Bus served a number of SOE missions on the west coast of Norway, of which *Antrum*, centred on Alesund half-way between Bergen and Trondheim, was typical. It supplied information about shipping and helped in sabotage operations mounted from Britain. A small resistance movement had been set up in Alesund in 1940 by Karl Aarsaether, who then had to escape to Britain. He was recruited and trained by SOE and at the end of 1941 returned to Norway with Fredrik Aaraas as W/T operator.

The preparatory work for their return was handled by Malcolm Munthe, who had joined SOE headquarters when he came back from Sweden. It included the provision of fake identity cards, the choice of cover, and the issue of weapons and provisions, including sugar, coffee, and tobacco. There was argument about Aaraas's identity. SOE saw him as a shipwright, which he objected to on the ground that the Gestapo could easily check shipyard personnel records. 'Student' would be safer. In the event he became a casual labourer who could find an anonymous job in the shipyards, where he would be well placed to gather information.

The pair left Shetland on 5 December 1941 on the *Siglaos* (Skipper Q/M Blystad) and landed near Alesund. A recent commando raid had increased German vigilance, which made life difficult. Survivors of the original resistance group were helpful but reluctant to have anything to do with a W/T operator and his dangerous job. Eventually one Finn Luth-Hansen, who had a small cottage 5 miles inland, agreed to take him in, and to be trained as a W/T operator. Although employed by the local telegraph office he was not an apt pupil, and when Aaraas returned to Britain much sooner than SOE had planned, his messages were still unreadable. Aarsaether returned a few days later. The debriefing of the pair reveals signs of the friction which was by no means unknown in SOE missions. Aaraas said the resistance group regarded Aarsaether as a mere courier. 'He is not admitted into their secrets more than necessary, as he is considered to be boastful, vain and is apt to drink.' Aarsaether said of Aaraas: 'He was not appreciated by the Alesund organization, as his accent proclaimed him to be a stranger.' With the departure of the two agents almost before they had arrived, SOE's representation in Alesund disappeared.

Munthe, still anxious to get a foothold in the district, planned to send in Aarsaether's brother Knut, who had just completed his W/T course. He was provided with a sports suit, a set of Norwegian underclothes, an automatic pistol, a thermos flask, a wrist watch, and Kr. 1,000. He found a passage on the fishing boat *Heland* (Skipper Sverre Roald) which was due to take 13 tons of stores to Alesund; and was instructed to complete Luth-Hansen's W/T training, and on no account to return before his pupil was well qualified.

There was now regular W/T communication with headquarters, although the mission was little more than a holding operation. Some of the messages were near panic. If the Allied invasion was delayed much longer, the Norwegian Home Front would collapse. The Alesund organization could not survive beyond the end of April. 'Send further 250 tommy guns and 100 Bren guns.' London urged patience, being well aware that there was no intention of invading Norway. Some more arms would be sent, but not the huge quantities asked for. Nevertheless

Knut Aarsaether kept surprisingly cheerful. In reporting that sixty men had been arrested and would probably be shot, he signed his telegram 'Harpist', having in mind a rumour that he himself had been shot. London replied that, when Alesund got too hot, he would be welcome in Britain where 'no harp-playing qualifications are needed to land'.

Knut came out at the end of March 1942, and yet again there was no SOE-trained agent in Alesund. At least *Antrum*'s W/T set had survived. It was hidden in a secret chamber in the home of Arthur Oerstenvik, a farmer and butcher. The farmhouse was several miles from the main road, difficult to see and impossible to approach unseen. In a cellar used as a washhouse there was stored some furniture, including a wooden cupboard standing in front of a home-made safe in which Oerstenvik kept money and valuables. To all outward appearance the safe was quite ordinary, the obvious place for a careful farmer to keep his prized possessions, but its back consisted of two steel plates sandwiching a 3-inch concrete slab indistinguishable from the walls of the cellar. The back could be removed to give access to a secret room walled with cement, and furnished with a bunk, electric fire, emergency rations, and all the paraphernalia of the W/T station. The aerial was erected through a ventilation pipe, which also provided two-way communication with the outside world by means of a tapping code. As the safe was half-way up the wall, entry into the secret room was awkward, but the ingenious farmer solved the problem by putting a hook in the ceiling of the cellar ostensibly to take a clothes line, but in reality to carry a rope to enable the W/T operator to swing, Tarzan-like, through the entrance. The safe back was then replaced and the cupboard repositioned.

Mission *Antrum* now enjoyed the benefit of the doubt which was present in many of SOE's enterprises at this stage. Although virtually nothing happened after Knut's departure, *Antrum* was deemed to be 'progressing very favourably'. Luth-Hansen continued to operate the W/T from the secret chamber, and trained an understudy, a member of the police reserve. Objectives for sabotage were earmarked—not an onerous task. The original escape organization lay dormant. Relations with the owners of country buses, which could be used for various purposes, were good. However, the main concern of the resistance seems to have been with financial matters. The Aarsaethers' mother was 'a very nervous lady' and had to be sent to Oslo when her sons were in Alesund—a costly business. Fishermen keeping their boats in repair for escape parties had to be financed. Boats acquired for refugees had to be paid for—Kr. 30,000 was not unusual. The rent of sheds to store arms involved considerable expense, as did the support of families of resistance workers who had been arrested.

During the next six months little more than token messages were exchanged with *Antrum*. In spite of the almost perfect security of the W/T station, it was claimed that signals must be kept to the minimum because Germans were billeted in a nearby farm. Anxiety about the presumed Allied invasion persisted: 'The people keenly impatient. Come soon!' London was unsympathetic and merely replied: 'Glad to hear you are taking care!' (an allusion to the paucity of W/T messages). 'You must not expect us to come soon!' It was gently hinted that the mission should now be more enterprising. Why not arrange W/T courses for new agents? Munthe became impatient at the lack of action and pointed out that *Antrum* had gone downhill ever since Knut Aarsaether had come back to Britain. The solution might be to send an operator to work independently of the existing station.

Karl Aarsaether was sent back on the *Bergholm* on 4 March 1943 with half a ton of stores; and at last *Antrum* began to function to London's satisfaction. Karl showed a new dynamism. The mission's earlier poor performance was blamed on Luth-Hansen, keen enough, 'but nervous when he is on the job, and also doesn't keep his mouth shut'. He had been got rid of 'in the nicest possible way'. The arms dump had been dispersed for greater security; there were proposals for the deployment of new agents; and information about shipping movements was provided for the use of the RAF.

While SOE's mission was beginning to find its feet, the indigenous resistance organization, which was almost totally inactive during 1943, was penetrated in December, when the Germans arrested twenty-two members. Aarsaether, who looked after those of the leaders who escaped, had no doubt that careless talk was responsible for the breakup of the organization. Although the group had been dormant for a long time, they could not stop talking about the business of resistance. The immediate reason for their downfall was penetration by a quisling who ingratiated himself with the fishermen on neighbouring islands, pretending to be a resistance leader looking for recruits. Most of the people he approached saw through him, but one accepted him as genuine and that was enough to demolish the whole group.

Aarsaether also looked after the seven survivors of the Shetland Bus *Bergholm* after an engagement with a German aircraft. His local 'collaborators' urged him in the interests of security to do no more than tell London that the survivors had got ashore; but when he realized they were the men who had brought him to Norway three weeks earlier, he thought he had to help them, in spite of the fact that German troops were searching for them all over the district, and a large reward was offered for information about them. He arranged for the seven to be taken to safety, and summoned a boat from Shetland to pick them up—all of which went without a hitch. The only worrying moment came when he was accosted by a German soldier, who wanted no more than a light for his cigarette.

When Aarsaether's health deteriorated at the end of 1943 he came out, bringing with him his fiancée, who had spent eight months in prison for helping refugees. They arrived in Shetland on the sub-chaser *Hitra* (Lieutenant Hauge) on 18 January 1944 after a long and nervewracking wait for very bad weather to subside. Karl was again succeeded by his brother Knut, who was given terms of reference calling for more action in the field. He was still required to provide intelligence, in particular about the movement of convoys and troops; and he was instructed to sabotage the ferry between Vaage and Nordvik, the car ferry *Geiranger*, and the salvage vessels in Alesund. Other tasks were to interrupt local telephone and telegraph communications, to encourage fishermen to deny their boats to the Germans, and to form a group to carry out anti-scorch in the event of a German withdrawal.

Knut, who landed on 18 February 1944, claimed complacently within a few weeks that his tasks 'were in the main completed'. The cutting of telecommunications had been assigned to the people concerned. The immobilization of the Vaage–Nordvik ferry had been passed to 'one or two safe men out there'. The *Geiranger* was still undergoing repairs. When they were completed, she would be immobilized again. Fishermen were being encouraged to put their boats out of action in the event of invasion by the Allies. Nothing had been done about the salvage vessels, since there were none in the region. Morale was not bad, but might be better. People could be more active against the Germans without getting into trouble; but for four years they had had a diet of passive resistance from the BBC, which should have attacked the lukewarm attitude of some patriots 'of whom there are not a few, instead of indulging in slimy flattery of the whole Home Front'. This had been resented by the active resistance movement.

Round about D-Day the intelligence required from *Antrum* was related to the Allied invasion of Europe: detailed information about troop movements, especially their destinations, although here there was little to tell as the local garrison was down to 400 men, mostly very old and very young. Three days before D-Day came an urgent request for news of ships, especially warships, moving south. *Antrum* was also used, as were the other SOE missions in Norway, to warn the people of the district not to be taken in by German instructions and false news; and to warn those who were compromised or were wanted by the Gestapo to lose no time in going underground.

For the remainder of the war *Antrum* was more concerned with paramilitary activities; and found itself in competition with the local contingent of the secret army *Milorg*.

Sabotage of shipping

Operation *Title*

In June 1942 the Admiralty placed an admiral on special duty to plan the destruction of the German battleship *Tirpitz*, then lying in Trondheimfjord; and the Royal Navy asked SOE, which had already provided intelligence about the battleship through Missions *Antrum* (Alesund) and *Lark* (Trondheim), to co-operate. The resultant operation was one of the most enterprising, most carefully planned, most daring, and most potentially valuable of the whole war. It failed, after coming within an ace of success.

The first plan was to ship two 'chariots'—in effect torpedoes 'ridden' by two men in diving suits—to an island off the Norwegian coast where they would be transferred to a local fishing boat to be towed submerged to the vicinity of the *Tirpitz*. There the 'charioteers' would don their diving suits, mount their electrically driven vehicles, and make for the battleship to place limpet mines at her most vulnerable points under water. This plan had to be abandoned because the fisherman earmarked to bring the chariots from the offshore island refused to co-operate. He was told he would have to do no more than tow an underwater container to Trondheim, but he would not take the risk since his aged parents depended on him. He would have been less willing had he known what was really in the wind. It proved impossible to find a substitute. All the other large boats with gear powerful enough to lift the chariots were owned by syndicates and it was unwise to bring several people into the arrangements.

The alternative—that a fishing boat should carry the chariots all the way from the Shetland Base to Trondheimfjord—was more hazardous since it meant passing the German controls and satisfying them that boat and crew had been going about their lawful business in Norway ever since the occupation, and that the cargo was properly documented. The boat chosen was the *Arthur*, which Leif Larsen had stolen in October 1941 when his own boat *Nordsjoen* was lost after a mine-laying expedition. She was a suitable choice since she was typical of the craft of the region and her three-man crew, P. O. Bjornoy (engineer), R. Strand (W/T operator) and J. R. Kalve (crewman) were equally typical. *Arthur*'s log-book had to be prepared recording all the trips she had supposedly made, duly rubber-stamped and signed by the appropriate port and Customs authorities; and identity cards and passes had to be provided for captain and crew. It was intended that *Lark* should supply samples of the various documents via Stockholm, but they took so long to arrive that Station XV, responsible for forgery, began to despair. Then 'like manna from heaven a ship arrived in Shetland from Norway . . . with a satisfactory set of papers to pass through the German control'. The papers were flown to London, where it took a week to make the special paper, cut type, and make the various rubber stamps. The papers were then flown back to Shetland to be filled in in consultation with the latest refugees who had up-to-the-minute information about the latest control formalities.

Arthur sailed from Lunna Voe about noon on 26 October 1942. The charioteers, Sub-Lieutenant Brewster, RNVR, Sergeant Craig, RE, and Seamen Brown and Evans, with their dressers, Seamen Tebb and Causter, whose job it was to help them into their diving suits, were hidden in a specially constructed priest's hole in the middle of the cargo of peat. The weather was very rough and in the first twenty-four hours the boat, with the two chariots lashed to her deck, could make no more than 3 knots. It was not until 8 a.m. on 29 October that they reached Hogoerne, where the chariots were to be submerged. First it was necessary to ensure that their batteries were fully charged, but after a few minutes the charging motor fell over and was damaged beyond repair. There was no alternative but to hope there was enough power in the batteries to complete the eventual short trip to the battleship.

At this point problems which no amount of planning can legislate for began to present themselves. Before the chariots had been submerged, but happily also before they had had their protective coverings removed, 'two fishermen came alongside and started talking to us. They were a bit suspicious as to what we had on deck and asked a lot of questions . . . It is very hard to get away from Norwegians who talk.' As soon as he had got rid of them, Larsen sank the chariots about 3 fathoms deep as the water was too shallow for the full length of the towing wires. The water was very clear and the chariots were quite visible from the surface. As soon as they were lowered, a fisherman appeared in a rowing boat. 'After looking all around he said, "What have you there?" I replied, "Something for demolishing mines." He then said, "It looks like a U-Boat," and I again repeated, "Something for demolishing mines." ' Larsen gave him some butter and coffee, and told him it was dangerous to talk. There was no guarantee that he would not report the curious objects *Arthur* was towing, but Larsen was pretty sure he would keep his mouth shut.

An unfavourable wind prevented *Arthur* from sailing until the afternoon of 30 October; and now the engine gave trouble and when they reached Hestviken Larsen went ashore to borrow the tools needed to take it down. It turned out that the cylinder was cracked and Bjornoy and the crew had to work all night making temporary repairs. By morning they reckoned it would get them to their destination, just. The voyage into Trondheimfjord started at 10 a.m. on 31 October. At Agdenes Larsen brought *Arthur* alongside the examination boat. The German control officer came on board, went into the after cabin and called for the ship's papers. Larsen handed him the crew list and the *Erlaubnisschein* to which were pinned copies of the order that no contact was to be made with England. He examined the papers carefully, and then remarked that he knew the German officer who had signed the *Erlaubnisschein*, adding, 'I know him from Bremen, but I did not know he was in Alesund.' This was a narrow escape. There could have been trouble if the control officer had known positively that the signatory had been in Bremen on the date when Station XV made him sign the paper in Alesund. However, all the ship's papers passed with flying colours, and the officer filled in a paper authorizing *Arthur* to proceed to Trondheim. His Norwegian was very poor and Larsen had difficulty in explaining what the cargo was, as the Norwegian for peat meant nothing to the German. When he was told it was something for burning, he was satisfied. The German seemed very anxious to start a conversation, but Larsen was equally anxious that he should leave as quickly as possible to minimize the danger that he might incriminate himself. Also, 'I was scared that the German officer would look down and see the chariots as the water was very clear'.

As *Arthur* went on her way down Trondheimfjord everyone was in great spirits. They had taken the most difficult hurdle in their stride, the sea was calm, the weather fine. The only incident was the passage of a

German destroyer, which drove the chariot teams from the deck into the shelter of the wheelhouse—the presence of six passengers who had not been declared to the control officer would take some explaining—and made it necessary to slow *Arthur* so that the wash from the German ship would not damage the chariots. Then disaster struck. In spite of the fine weather, about fifteen minutes after passing the Rodberget Light 'we suddenly ran into two fairly large waves. The boat pitched and we could feel a drag on the chariots, and a second afterwards one of them hit the propeller.' Larsen had no doubt from 'the feel' of *Arthur* that the chariots had gone. This was later confirmed when Evans, one of the charioteers, put on his diving suit and went down to see what had happened. The towing wires were still attached to the boat, and at their ends were the lead weights used to keep the chariots well below the surface. But that was all.

There was nothing for it now but to try to carry out one of the escape plans which had been prepared in consultation with *Lark*. The seacocks were opened and *Arthur* left to sink while the ten men rowed ashore. They split into two groups for the journey to the Swedish border: Larsen with Kalve, Craig, Tebb and Evans in one, and Brewster with Bjornoy, Strand, Causter and Brown in the other. The latter party reached Sweden without incident, but, after five days in which they were looked after by friendly farmers, Larsen's party ran into trouble within a short distance of the border. They were refused shelter for the night at four neighbouring farms, at one of which the farmer's wife was particularly unpleasant. They became suspicious, thinking they might be reported to the police, and decided to move on quickly. But before they could leave the area they were challenged by two policemen. According to Kalve's account:

> When we heard the order to halt we all stopped and Larsen, knowing the British had pistols, turned round and asked them in English to draw them. However, owing to their cumbersome clothes they could not draw their pistols fast enough and this gave the two men time to walk right up to our party. Larsen and I were covered at close range . . . We were ordered to put our hands up and accompany the two men . . . We were told to follow the two men and Larsen started arguing. This gave one of the British time to draw his pistol. He fired three times and hit one of the men.

The other policeman returned the fire, wounding Evans. In Larsen's words: 'Then everybody started to run. There was a fence just beside the road. Some of us started to jump over it. Tebb and I ran down the road a little further, crossed the field and managed to reach the river . . . All firing had ceased and everything was quiet.' It was decided there was no hope of rescuing the wounded Evans, even if he were still alive; and by next day the remainder of the party crossed into Sweden where they were well received. For some time it was assumed that Evans must have died of his wounds, but the truth was revealed six months later by one of the policemen who had accosted the *Title* party, who, although a quisling, had found it necessary to escape to Sweden.

Larsen's suspicion that they had been betrayed was confirmed. The unpleasant farmer's wife had directed them to a farm nearer the police post, and had then telephoned the police to look out for five suspicious-looking men. Two policemen went to the place indicated and in the pitch darkness heard the men approaching. As they were marching them to the police station, one turned and shot one of the policemen, who before he fell unconscious emptied his revolver in the direction of the prisoners. When he came to he was alone except for a wounded man, who was groaning and trying to sit up. Afraid that the man might be armed, the policeman immediately left, half walking and half crawling. Back at the police station he gave the alarm, before being taken to hospital.

Next day the German police were out in force. They found four sets of tracks in the snow heading for the border, three together and one—Kalve's—on its own; and they concluded that the four men were now beyond their reach. The unfortunate Evans had managed, despite the wound in his thigh, to drag himself to a hut, but the police had no difficulty in following his tracks. He was wounded again in resisting arrest. He gave nothing away under interrogation, but when the diving suits were found in the *Arthur*, which instead of sinking as planned had drifted aground in shallow water, it was obvious what part he had played. He was shot on 23 January 1943, 'in accordance with the Führer's orders' that saboteurs should be summarily executed.

Events had made it impossible for Larsen and his men to keep the rendezvous with *Lark* as part of the arrangements to get them safely to Sweden; and *Lark* could only speculate as to what had happened. Clearly something had happened, as twenty-five students were arrested for no apparent reason, and it was guessed it might have something to do with the *Tirpitz*. *Lark*'s first message raised great hopes in London:

Congratulations. *Tirpitz* has a list and oil alongside. Repair vessel ordered in. Extent and nature of damage not yet known. Your boat unfortunately sunk in such shallow water that she is visible. Hundreds of Germans have combed Frosta for several days. Crew not captured and am confident specialists [the charioteers] are not either.

On the strength of this an excited telegram was sent off to the Shetland Base where the crews were to be told that *Arthur* with their four compatriots had carried out a most successful operation off the Norwegian coast. *Arthur* had been scuttled and the crew were safe. Aerial reconnaissance on 1 November showed a mile-long oil slick from *Tirpitz*'s stern and there were indications ('even allowing for wishful thinking') that the ship was lower in the water than usual. London sent an equally euphoric telegram to *Lark*. Congratulations were due to him and his people. News of the extent of the damage would be welcome. Alas, the next news was that there had been no damage—the damage *Lark*'s man had reported had been caused before *Tirpitz* arrived at her present anchorage. In the final inquest it was established that the reason the chariots had been lost was that the tow ropes had been tied to the chariot hand rails which had failed to take the strain of the heavy seas. It may be a matter of regret that, in an operation planned to the most minute detail, no one had foreseen this crucial flaw in the arrangements.

Operation *Carhampton*

In March 1942 Ottar Grundvig, recently escaped from Norway, proposed the capture of Norwegian ships from some small fjord where the German presence was slight. He believed Norwegians deprecated the exploitation of their merchant navy, and if some vessels were transferred to the Allies it would boost morale. 'It is well known that the spies in Norway and especially those who disclose shipping movements are sentenced to be shot after first having been tortured, and one cannot expect they will keep on working when they see little or no result from their work.' His ideas were passed on to SOE, which put them to Odd Starheim, one of the first Norwegians to escape to Britain. He had arrived in Aberdeen in a small boat with two companions in August 1940, and with three other recent arrivals became the first Norwegian SOE-trained agents.

In January 1941 Starheim landed from a submarine on the south-west coast of Norway, despite being far from well at the time. He spent the next five months organizing resistance in Agder, and then returned to Britain by fishing boat. In January 1942 he parachuted into Vest Agder as *Cheese*, accompanied by Sergeant Fasting (*Biscuit*), with only rucksacks and a W/T set. When he visited a contact in Oslo in February the Gestapo surrounded the house, but he exploited 'the time-honoured trick of visiting the bathroom', locked the door, and jumped from the first-floor window, damaging a foot on the frozen ground. He thumbed a lift and got away; but for the time being Norway was too hot for him. He decided to return to Britain in style in a stolen coastal steamer. With five friends he boarded the 600-ton *Galtesund* as fare-paying passengers, and, when clear of land, held up captain and crew and prevailed on them to make for Aberdeen. There were only twenty-one on board—fortunately a party of German soldiers had missed the boat at her last port of call, otherwise the piracy might not have gone so smoothly. So confident had Starheim been of success that he instructed a W/T operator staying behind to signal to London for an escort. This was the second time he had brought a ship from Norway and, when it was proposed to steal a whole convoy, he was the obvious choice for the job.

He accepted with alacrity.

I am prepared to take responsibility for the operation. I am a navigator and an officer in the army and have experience in automatic weapons. I know the conditions round Flekkefjord and the coast there very well. It was at this place that I planned and carried out the capture of the *Galtesund*.

He prepared a plan down to the last detail. The best place for the attack would be off Abelsnes in Flekkefjord. Small convoys, about four ships, often spent the night there, and the fifteen-man guard post could be eliminated by half their number. Since the post reported to Flekkefjord every two hours, the attack must be carried out immediately after a routine 'all's well' call had gone through. Ten men would be needed to board each of the presumed four ships, armed with tommy guns, pistols, knives, and truncheons. Each man would carry rope round his waist to bind prisoners, but this provision was abandoned when it was pointed out that it was against international law. (There was no objection to shooting an enemy before he was made prisoner, but there was objection to binding him after capture.)

Each party of ten would have a boat with muffled oars, a boat-hook, and a rope ladder to board their ship. When the shore guard had been silently disposed of, boarding parties, ten-strong for four ships, larger if there were fewer ships, would with clockwork precision overpower any guards on board, take over the engine room, stokehold, and W/T room, confine captain and crew in a cabin from which they could not signal to the shore, cut the anchor chain with plastic explosive, and at midnight, exactly thirty minutes after the first man stepped on board, sail out the convoy. By morning it would have travelled 40 miles and soon afterwards would be looking to the RAF for protection.

It was not until November 1942 that *Carhampton*, as the operation was code-named, got under way. A force of forty, precisely the number Starheim had specified, set off for Abelsnes on 15 November in the whale-catcher *Bodø*: twelve from the Royal Norwegian Navy, and twenty-eight from NIC 1 who had undergone intensive training for the operation. The rough sea ruled out a landing and a second attempt made shortly afterwards was thwarted by bright moonlight. *Bodø* made a third attempt, and after an uneventful voyage in good weather landed the party at Televik, some 7 miles from Abelsnes on 3 January 1943. On the return trip, *Bodø* struck a mine and sank almost instantly. The only survivors of the crew of thirty-five were Lieutenant William Johannson, RNN, and Petty Officer Fritzholm. There had been time to launch only one raft, on which five men found themselves. Two died on the raft, and a third on the rescue launch which arrived eventually.

The *Carhampton* party established themselves in an isolated empty house and shed at Televik. Sergeants Winge and van der Hagen set off to make contact with the resistance organization at Flekkefjord but were driven back by a heavy snowstorm. After a few hours' sleep they tried again. They got to their contact's house as he was saying goodbye to his family, before leaving for Sweden to escape the attentions of the Gestapo; but when he heard what was planned he agreed to stay and help. Sergeants Winge and Aubert were detailed to occupy a hut on the west side of Flekkefjord from which they could reconnoitre the guard post at Abelsnes, locate the best place to cut the telephone wires, and report when a convoy arrived. In the next week the only event of note was the appearance of three fishermen at the group's headquarters, who had to be kept prisoner in the interest of security.

On 10 January Aubert reported a three-ship convoy, but he knew little about ships and was uncertain about their size and nationality. Preparations for attack were made immediately. At first it had been hoped that the whole party would go by road in two trucks, but only one was available. Sergeants Winge and Hansen used this to take their squads to Abelsnes where the telephone lines would be cut. The other two squads under Starheim would go by boat to a creek about twenty minutes' walk from Abelsnes, where they would rendezvous with

Aubert, who knew the district well, and by this time would have led Winge's squad to the guard post.

Everything went remarkably wrong. Although the telephone wires were cut, Aubert left Winge some distance from the guard post, and it was never found. He also failed to rendezvous with Starheim who made his way to Abelsnes unaided. Winge's squad

> had maps and sketches with them and ought to have found their way without any help . . . I might mention also that it was moonlight, so it was very easy to find one's bearings. We were very disappointed when we heard that the sentry posts had not been dealt with, and with that disappeared also our dream of walking down Piccadilly in a week's time.

Thus van der Hagen. Starheim decided to go for the convoy without neutralizing the guard post, but it took so long to get the boats ready that, even if the ships had been taken over without a hitch, it would have been daylight when they passed Hidra and they would have been at the mercy of the German batteries there. Starheim ordered everyone back to Televik.

The retreat was also something of a fiasco. In van der Hagen's account:

> It was about a 12 km. [7-mile] march, but the condition of so many of the men was so bad that they went into huts on the way. Lieutenant Starheim, who had less training than any of us, had to go backwards and forwards among the boys to get them to keep going. There were at least 5 km. [3 miles] between the first and last squad. Most of them got through at last and those who had gone into the huts were fetched next day by motor car.

Van der Hagen attributed their poor performance to 'nervousness' for they had had no serious exertion for some days and should easily have covered 7 miles in three hours.

Next day van der Hagen went into Abelsnes to find out if the abortive attempt had alerted the Germans and was horrified to find that the team cutting the telephone wires had left hand grenades and English mittens behind. Sentries in Abelsnes were being doubled, and sixty men were being drafted in from the *Feldgendarmerie* in Arendal. Worse, there had been reports that the *Carhampton* team had been seen on the Televik road. It was imperative to move out right away, but a local resistance man reported that the convoy had been replaced by a much more worthwhile target—four ships of 6,000–7,000 tons. This was tempting and, despite the increased danger, Starheim decided to try again. The whole party, having travelled most of the way by boat, reached Abelsnes just before midnight. This time Starheim's own group set out to attack the guard post, but, when they were within a few feet of the sentries, one turned and fired. As there was now no hope of disposing of the guard silently—one of the key elements in the plan—battle was joined with grenades and pistols.

> Lieutenant Starheim wanted us to storm the hut but when we began to collect the men there were only 3 or 4 of them left. The others had taken the road up to the moors, so it was decided we should get down to the other squads as quickly as possible, get the rowing boats ready and cross the Feddefjord.

This was the only line of retreat. If the original plan of moving towards Flekkefjord had been followed, they would have walked into the arms of a German force coming from that direction. The trip was exciting as the fjord was swept by searchlights from ships and shore, and the scene was also illuminated by Very lights. The boats were sunk on the far shore of the fjord, and the party began a long trek through deep snow in the most trying conditions, in which the health of many suffered, often due to frostbite. They had a great deal of help from the local people, including two doctors, and after many days reached Eiken, 30 miles from Abelsnes as the crow flies, but much longer by the hazardous route they were forced to take.

They still had their W/T and could keep in touch with London. Having failed in his main objective, Starheim was all the more anxious to accomplish something. He put several proposals to headquarters, including the demolition of the mines at Knaben, the capture of boats at Farsund, and an attack on the German garrison at Lynsvaagn in conjunction with a force from the Shetland Base, all of which were turned down. A plan to capture a 10,000-ton ore carrier at Regefjord, which was approved by London, was abandoned when it was found that security was impossibly tight. Other plans were considered, the final being the capture of the coastal steamer *Tromoysund*, a repetition of the *Galtesund* ploy, which London endorsed. To carry it out it was necessary to get the whole party back to the coast, which meant a train journey from Sirnes to Heskestad.

At Sirnes station they dumped their obviously British rucksacks on the platform and stood as far away from them as possible. They bought tickets to different destinations, all of which would enable them to leave the train at Heskestad. Their first narrow escape came when they tried to board a German troop train, but were restrained by the station master. Their shabby dress was looked at askance by many of their fellow passengers, but no one did more than look. On the train, which they boarded in twos and threes in different carriages, Sergeant Vilnes

> heard a voice that I knew well behind me, so I pulled my hat well over my eyes and looked round. Sure enough it was a chap from Alesund who knows me very well, but who fortunately did not notice me. I had to move to a seat next to a young woman, with two German NCOs facing us. Since I had an overcoat over my battledress, I had to button it tightly at the neck to hide my uniform, and my paper bag with 8 hand grenades inside had to be held carefully to keep it from falling apart.

Half the party was to capture the steamer before she left Regefjord on 28 February. The other half would join fishing boats and sail them to Britain. Rolf Olsen would board the steamer at Flekkefjord to look for hidden snags on the journey to Regefjord; and, if all went well, he would become second in command to Winge. If he failed to get on board at Flekkefjord, Martinsen would be second in command. Otherwise Starheim and Martinsen would leave *Tromoysund* as soon as *Carhampton* was in control, to organize the departure of the rest of the group. The boarding party, wearing the most respectable clothes they could muster, walked to the pier in twos and threes at five-minute intervals. When the third and final 'all visitors ashore!' bell sounded Starheim, Winge, and Martinsen mounted the gangway to overpower the guard, and the rest of the party waiting on the pier stormed after them. Corporal Bunch had to cover the telephone exchange to ensure that no SOS was sent, and then join the ship as she was leaving. Corporal Endresen watched progress from the shore, so that, if all went well, a signal could be sent to London. He heard some shooting before the ship left the narrow harbour with some difficulty, making poor headway as she moved down the coast against a choppy sea; and he saw a man 'in an overcoat and light coloured cap which answered to Lieutenant Starheim's description going up the road away from the pier'.

It now seemed to the shore group, led by Vilnes, that everything was under control; but then neither Starheim nor Martinsen returned. There must have been a change of plan. While Vilnes waited anxiously for his leader, three cars arrived at the barn next to the one where they were hiding, and disgorged ten men who made a thorough search of the barn and surrounding area. It was obvious that their own hiding place must be the next port of call, so they at once moved to the hills, where they remained for a week. On 4 March Vilnes sent Endresen down to

Egersund to negotiate for a fishing boat to take the whole party to Britain. Three brothers agreed to help and on 8 March nineteen of the *Carhampton* team were smuggled on board the *Ann Elizabeth*, which reached Hartlepool five days later after a very rough voyage. Four others volunteered to remain behind in a mountain hideout with the W/T set; and one sick man and an escort had earlier gone out through Sweden.

The other sixteen members were on the *Tromoysund*, which was less fortunate than the *Galtesund*. One day out from Regefjord she was shadowed by German aircraft, bombed and sunk. Five survivors were sighted by the RAF on a raft and in a lifeboat, and a destroyer was sent to their rescue, but it had to be recalled when they drifted eastwards into enemy minefields. That Starheim had remained on board, whatever the reason, was established when his body was later washed ashore, and identified by his wristwatch.

After reading the account of *Carhampton*, Selborne commented: 'An amazing story. I cannot help feeling that these men's lives were risked on a plan that was not worked out like *Gunnerside*. But it is all in the day's work.' Perhaps this was an ungenerous summing up. The original plan was all that it could have been, but the conception was unsound. The Germans must have been waiting for a repetition of the *Galtesund* operation, and, when the *Tromoysund* failed to reach her next port of call, it was obvious to them what had happened, if indeed they had not been alerted by the shooting on board; and she had no chance of getting beyond the range of enemy aircraft in the hours of darkness available. The same must have been true of any convoy that *Carhampton* took over. The Norwegian Section found it difficult to assess the operation, which had cost 100 lives, including sixteen SOE men. Adamson stressed the useful experience gained in the field. Wilson was critical of

> the lack of proper and exact reconnaissance coupled with the careful sifting of information received. Discipline was good, but excessive use was possibly made of the Norwegian democratic principle of consulting everyone before a decision is taken.

He also shrewdly pointed out that the dramatic success of the *Galtesund* operation had blinded everyone to the greater possibilities that would have been opened up if *Carhampton* had concentrated on limpeting the convoys it encountered. A greater tonnage could have been sunk more easily than commandeered.

Operation *Mardonius*

This ambitious attack on shipping in Oslo harbour was planned by Max Manus, whose colourful career had taken him to South America, where he was 'a sailor and to some extent arms smuggler, in Venezuela and Colombia' and was briefly imprisoned in Chile; to Finland to fight against the Russians; and back to Norway in the vain attempt to keep out the Germans. He then organized resistance in Oslo and carried on a propaganda campaign through underground newspapers. He was arrested by the Gestapo, dived through a first-floor window, found himself in hospital under German guard, and a month later, when he was deemed fit enough to be interrogated (with all that that meant), was rescued by his companions who smuggled a fishing line to him in hospital, enabling him to pull up a rope down which he slid into their waiting arms, although he still had only one sound hand. After convalescing in a hide-out (with the luxury of a gramophone and a single record, *The Greatest Mistake of my Life*), he crossed into Sweden with Andreas Aubert, and finally reached Britain after one of those not unusual wartime longer-than-the-crow-flies journeys through Finland, Russia, Romania, Turkey, Egypt, South Africa, Trinidad, the United States, and Canada. In Britain he joined the Linge Company and went through the SOE training courses.

The plan was that Manus and Gregers Gram, who had also taken part in resistance work in Oslo, should drop near the capital to recruit expert canoeists, instruct them in ship sabotage, and attack ships in Oslo harbour. When their training in Britain was complete, they dropped in the night of 12–13 March 1943 with forty limpet mines, three Sten guns, pistols, and a substantial quantity of food. Their equipment also included fifteen suicide pills for the use of the team if it was captured. Before take-off they were disconcerted to observe that the pilot's map showed neither the road, lake, nor hill which were their chosen landmarks. They hastily supplemented it with a sketch map of the area.

Manus wanted to carry out the attack so that no suspicion rested on patriots who would suffer German reprisals. The BBC could help by putting out a story that a party of Britons had been using a new sort of mine; or perhaps other agents could be dropped near Oslo and then make for Sweden, having left obvious traces of their presence. In fact neither of these precautions was taken. The flight, bitterly cold, was without incident; but there was a problem as they left the aircraft. Gram's parachute was momentarily caught in Manus's static line. They landed in difficult terrain at the south-east end of Tonevand Lake. Manus saw he was heading for rocky ground but managed to divert himself into a marsh. Gram landed on a hillside sloping down to Steinsjo Lake. He had almost reached the ground before regaining full control of his parachute but made a safe landing in a tree. The pair, both convinced that the other must be injured, had to find each other in the dark, as the moon had disappeared; and after half an hour they met. They found a sheltered place, ate their emergency rations, and settled down for the night using a parachute as wind-breaker. Next morning they found all their containers.

They now had to make a permanent base. They pitched their tent in thick brushwood about 150 yards from Steinsjo Lake, hiding it so well that they later had difficulty in finding it. Manus had a bout of flu and Gram also had a fever. For a week both were very ill, and decided they must find a house where they could recover. They got in touch with a friend whose mother was in hospital, and who agreed to put them up. They had difficulty in finding recruits. One man they approached opted out on health grounds, then, troubled by conscience, said he would join, and then sent a message that he was ill again. Manus blamed the 'wait for the invasion' apathy for the lack of enthusiasm. 'We went to types we were fairly certain would help us; and we even believed—rather naïvely—that they would be glad of the opportunity. We had calculated on mobilizing up to 20 men, but we had the hell of a job to raise 3.' Even the 'sportsmen and politically-conscious anti-Nazis' made excuses. They had to think of their wives, it was foolish to take risks now, far better wait for The Day. 'Two tried to talk us out of the whole business, said at the best we would be shot, at the worst there would be a state of emergency.' The three volunteers were Halvor Haddeland, Einar Riis-Johansen, and Sigurd Jacobsen.

They were given instruction in Sten gun, revolver, fighting knife, and truncheon; and, as even they were sceptical about success, Manus arranged a dress rehearsal on a hulk which convinced them that they could succeed. Weeks passed without a worthwhile target coming into the harbour; and, as the nights became shorter, there was a noticeable drop in the enthusiasm of the local men. Manus offered them the chance of withdrawing, which was good psychology. All three immediately confirmed that they wanted to go ahead, provided the attack took place before Easter, after which the nights would be much too short. Almost immediately four ships came in, to provide reasonable targets.

The plan was to set up a base on the Island of Blekoya, where equipment could be hidden in advance, and which canoeists could visit quite openly. Manus and Riis-Johansen were assigned two targets—the *Tugela*, and the refrigerator ship *Windrich von Kniprode*. Gram and Haddeland had a larger choice in the central and western parts of the harbour. All four wore wind jackets over their uniform. As they paddled towards Blekoya the sentry on a watch tower scrutinized them through field glasses, but decided they were harmless. As they waited for darkness the clouds began to clear and, in Manus's words, 'great ugly stars were beginning to appear'. A patrol boat began to steam round the harbour focusing its searchlight on quays, ships, shore, and islands. Manus ordered his team to take up a defensive position, but the boat disappeared up the fjord. It was now or never, since, if there was no attack in the harbour, it seemed inevitable that the sabotage which Jacobsen was due to carry out in the Akers shipyard would be attributed to local people, which Manus wanted to avoid.

Gram and Haddeland launched their canoe shortly after midnight. They had made up four charges, two of four limpets each, and two single limpets. To ensure evidence of some activity in the harbour, they planted a single limpet on an unguarded oil lighter. They then moved cautiously towards the *Ortelsburg* of Hamburg, paddling with great care to avoid both noise and phosphorescence. They had no difficulty in attaching their limpets, the only man visible on the ship being an anti-aircraft look-out who was watching the sky.

Cautiously they made their way to the next target, now believed to have been the Norwegian vessel *Sarpforss*. As they worked their way alongside, they heard to their dismay footsteps across the deck just above them. A man appeared at the rail but his subsequent actions showed that he was unaware of the presence of the canoe below. His purpose was merely to relieve himself, which he did to such an effect that the canoe and its occupants received a most unsavoury shower. He then walked away and the limpet men coolly moved aft to the engine room, outside which the 4-limpet charge was carefully placed.

As they paddled to place their final limpet on a gas container serving a beacon light they saw the phosphorescent wake of the other canoe, which was at one time caught in the floodlights of Manus's first target, the *Tugela*.

Manus and Riis-Johansen left Blekoya half an hour after the other canoe. Their task was more difficult, since they had to contend with the men loading the *Tugela*.

To come out from under the wharf towards the *Tugela* was like coming into bright sunshine. Work was going full blast, and it was difficult to know what to do. The fact that to go back was now as hard as to go on caused Manus to decide on a bold stroke. They therefore shot out towards the stern of the ship, into full view. Manus thinks that the workmen were either blind, or thought they were seeing things.

Although they had intended to go on to the *Windrich von Kniprode*, it meant paddling across a large floodlit area and Manus decided not to push their luck further. The pair waited until the *Tugela* winches were at their noisiest and then dashed across the lighted area to the safety of the shore.

They had all used long-delay fuses, and next day Haddeland took a camera to the harbour to photograph their handiwork. As he arrived, he heard a loud explosion and saw a column of water shoot up beside the *Ortelsburg*, which listed and sank in a matter of minutes. Moments later there was a second explosion—a limpet which Manus had left on Blekoya to cause a diversion. Germans were running about all over the area, many of them undressed. Next, the charge on the oil lighter exploded. A man in a row boat made towards the lighter, but beat a hasty retreat when the beacon's gas tank blew up within a few yards.

The best item of the comedy was still to come. The workmen and the German guards at Gronlia had all swarmed aboard the *Tugela* and lined the rails and upper works to watch the fun. Haddeland himself was standing quite near. Suddenly the charge laid under *Tugela*'s stern exploded and within seconds the ship was purged of her spectators. Haddeland thought it wisest to leave the scene at this stage.

The activity planned for the shipyard had failed. Jacobsen had limpeted a minesweeper and a 10,000-ton tanker, the *Taiwan*; but the charges failed to explode. The limpets were spotted next day in the clear water, and were safely removed by the Germans. One which had not exploded on the *Tugela* was also removed as was that on the *Sarpforss*. *Mardonius* had been less than perfectly successful. The team left for Sweden immediately after the attack, and crossed the border, without noticing it, less than a week later.

To attribute the operation to external forces, Gram left a letter on Blekoya addressed to Admiral Doenitz, and pinned to a tree with a commando knife:

'Wherever the British attempt to operate on the Continent they will fail. Our glorious Wehrmacht will always be on guard and will destroy them.'
A. Hitler, Corporal III Reich, Berlin, 1941

'That's what you think, brother!'
Tommy Atkins, Corporal, III Submarine Commando
Oslo, 1943
PS *Sieg Heil*!

There followed a list of the places where Commandos had so far landed. The effusion was written on the back of a label from a British First Aid outfit, so that there could be no doubt about its provenance.

Operation *Bundle*

Manus and Gram returned to the field in October 1943 to spread subversive propaganda among German troops (see Chapter 10), and carry out further sabotage against shipping. Their first attempts—two attacks on 23 December 1943—failed because of increased enemy vigilance. In February 1944 things went rather better. The RAF had dropped mines in Oslo's inner harbour and the Germans had moved shipping out while they looked for unexploded mines. A newly built 600-ton ship, about to undergo her trials, was moved to a vulnerable position, where, unlike the other vessels, she had only a single Norwegian guard on board. The crew due to take the boat over was still living ashore; but a few Norwegian workmen went on board daily to put finishing touches to the vessel.

Gram planned to take advantage of the poor security. He would go on board dressed as a workman in broad daylight, plant a limpet, and walk off again. This is exactly what happened. It took two men to carry the limpet and ancillary gear, so he enlisted the help of Einar Juden:

> We started in the middle of the night (07.00 hours) and went down to the harbour loaded with limpets and weapons. We were delayed a bit by the German navy (which crawled all over the ship) and all the errand boys in Oslo who seemed to have some sort of a general meeting exactly at our quay. Instead of going on board at 8.00 o'clock as planned we drank a few basins of ersatz coffee in the nearby cafés till the coast was more or less clear.

They had to wait until 12.30 for the coast to clear. Then: '[The plan] was very simple and undramatic. It was just to board the ship by the gangway, walk into a deckhouse with the hope that it was empty, walk out again, climb over the rail, and down the side of the ship, place the blessed thing and leave—with long steps I admit.' The single limpet with an anti-removal device and a 20-hour delay exploded after eighteen hours, probably, Gram thought, because it had been so long in warm cafés. It had been timed to go off after the workmen had left. The boat sank in shallow water, but hopes that the boilers would burst when the cold water rushed in were not realized. However, as Gram records at the end of his report: 'The German reaction was not enthusiastic.'

Operations *Vestige*

Of the series of eight *Vestige* operations using kayaks or folboats to limpet shipping on the Norwegian coast, only two paid off. *Vestige I* (a three-man team led by Corporal Harald Svindseth with Sergeant Ragnar Ulstein as W/T operator and Corporal Nils Fjeld) landed near Gulen from an MTB on 3 September 1943. The beach was very unwelcoming, with little cover, so the team moved into the hills to a temporary hiding place to reconnoitre the area. Next day the Germans made a thorough search of the beaches and nearby islands, and a spotter plane flew low over their hide-out. Neither they nor their three kayaks were seen. As the beach was useless for launching small boats, they enlisted the help of a patriot farmer who took them to a safe house in North Gulen.

There was plenty of shipping but at first the moon was too bright for an attack. On 23 September it was dark enough to contemplate limpeting the most promising target, the 2,700-ton *Hertmut*, a modern refrigerated vessel escorted by a torpedo boat. As one man would have a better chance of success, the three drew lots and Svindseth won. They decided to use limpets with a 26-hour delay, to give the ship time to get well out to sea before the explosion, to make her destruction the more certain and to lead the Germans to suspect a submarine attack—as well as to leave the trio a good chance of a repeat performance.

Svindseth paddled out to the ship with three limpets between his legs. The other two followed in a rowing boat to back him up.

> When I got to within 20 metres [about 20 yards] of the ship, I discovered that the torpedo boat was lying along her port side. I worked my way along the starboard side and attached my holdfast to the hull. The kayak remained fairly steady. Taking one limpet at a time I broke the ampoules, screwed off the anti-removal fuse plates, put the limpets on placer rods and put them into the water as far down as the rods would reach, after which I guided them to the ship's side. While I was doing this no sound could be heard from the merchant ship.

The limpets exploded a mere forty minutes after they had been planted—a whole day too soon—just as Svindseth rejoined the others. The captain tried to beach the ship and would have been defeated by the sheer cliffs, had it not been for a helpful fisherman who hastened to the scene and guided the stricken vessel to the only shallow water for miles around. Within a week a salvage vessel had refloated her and towed her away for repairs. The belief that the explosion would be attributed to a submarine's torpedo was confirmed when her escort dropped many depth charges. As it was now too dangerous for *Vestige I* to remain at Gulen, they left the area. The leader's report gives some idea of the hardships agents faced on missions of this sort. They set off, without any rest:

> It was a steep climb up from North Gulen. The rain and hail soon made our rucksacks wet through. I suppose we were carrying about 35 kg. [80 lb.] apiece. The climb took us four and a half hours. At 12.30 we had a rest in a fir wood about a kilometre [about 1,000 yards] from the head of North Gulen. There we had our first meal since the previous evening. We realized at this point that our journey was going to take at least two days since the going was extremely hard.

It was five days before they reached their safe house; and, before they could signal headquarters to arrange a rendezvous, their host had to retrieve their hidden W/T. They contacted London on 17 October, and stood by on four occasions before an MTB arrived to pick them up.

The only other successful operation in the series was *Vestige III* (Corporals S. Synnes and H. Hoel) which *Antrum* received. Convoys were no longer visiting Alesund, unless to discharge cargo, so there were few suitable targets. Eventually the 7,000-ton coal ship *Jantze Fritzen* presented herself, and after a strenuous ninety-minute paddle *Vestige III* planted six limpets in pairs, which exploded before the pair could get back to the shore. There was much searchlight and patrol activity, so the men sank their kayaks, intending to return to *Antrum* overland. They spent one night in a barn, buried in hay with their equipment, unnoticed by the farmer who gave a neighbour a graphic account of their exploit. There had been sabotage on board the German ship at the Bunker Station, where the staff and several of the crew had been arrested on suspicion. Shortly after the explosion three tugs and a salvage boat had arrived to keep the collier afloat and had towed her into the inner harbour. Later she was towed to Bergen for substantial repairs.

The rest of the journey to *Antrum* was uneventful as the weather was bad and the road deserted. The members of *Vestige III* were picked up by a boat from Shetland on 17 September 1943.

Operation *Guidance*

An attack against a floating dock in Bergen harbour in November 1943 by four Welman* one-man submarines failed. The leading craft was manned by Sergeant Pedersen, who had taken part in the limpeting of the *Nordfahrt* at Orkla (see Chapter 9). As he approached the target area through fine rain and fog patches, he saw a small boat heading for him, which forced him to dive. He proceeded under water for fifteen minutes, when he reckoned he would be inside the German watch boats; but when he surfaced to get his bearings, he had to open the hatch and stand up. He was within 50 yards of a minesweeper, which turned its searchlight on him and opened fire with a 20 mm gun before he could dive again. As the Welman was hit, he flooded it, hoping to send it beyond the reach of the Germans, and jumped into the rubber dinghy in which they approached. He guessed that, even if he had managed to submerge, depth charges would have been dropped against his craft, which would probably have disabled the other three—which returned unharmed to the mother ship when the operation was called off. In fact his Welman did not sink fast enough and the Germans retrieved it. Pedersen found himself a prisoner of war.

* The Welman was 19 feet long, powered by an electric motor, and carried a 600 lb. bomb.

6

The secret army: Denmark

RONALD Turnbull arrived in Stockholm in March 1941 to take up his duties as SO 2's man responsible for Denmark. The suspension of the air service to Sweden had forced him to travel many thousands of miles and for many weeks through South Africa, Turkey, and Russia in order to reach Stockholm. He was immediately joined by Alfred Christensen, who was to be his lieutenant in Gothenburg with the cover of Vice-Consul, and who made the trip in a matter of hours, thanks to the resumed air service. At first Turnbull concentrated on propaganda and, in May, London, anxious to see more positive action, became restive. Hambro recorded: 'Turnbull wants jerking up. He thinks he is in the Ministry of Information. What is he doing about SO 2 work?' Lieutenant-Commander R. C. Hollingworth, head of the Danish Section at headquarters in London, duly jerked him up. He told him his many ideas for a propaganda campaign were admirable, but 'it should not be forgotten that your chief aim is the subversive'. Little or nothing had been done to create a special operations organization in Denmark, and there must be some visible progress in the next months.

Turnbull admitted that so far he had done little, and said he would make no excuses. He had thought he was expected to concentrate on propaganda, and go easy on sabotage, except for opening up lines of communication with Denmark. The balance would now change, although he still had to overcome the opposition of Victor Mallet, the British Minister in Stockholm, who had been anything but co-operative at the time of the Rickman débâcle and was still a problem. Turnbull had recently been called in by the Minister and 'harangued . . . on the uselessness of SO 2 work as a whole and with reference to Denmark in particular. This sort of thing does not help. Instead of finding at least sympathetic tolerance here one has to face actual and determined opposition.' The Military Attaché, who was in favour of special operations had been deliberately excluded from the meeting. Turnbull added that he was frequently 'assaulted' by other members of the Legation and asked 'what the hell I think SO is doing, and why the hell it is doing anything anyway. Well, as a miserable little Assistant Press Attaché [his cover] I am hardly in a position to answer back in the terms I should like to use.' Mallet was insisting that he should help with the work of the Press Department, even on Sundays, ostensibly to support his cover, but in reality to leave less time for his SOE duties. London accepted that the Minister wanted to keep SOE at arm's length, but reassured Turnbull: 'We should not allow subversive organization in Denmark to be hampered by fear of treading on the Minister's diplomatic toes.' Hambro thought the time might be ripe to 'start a row in Denmark'—perhaps by limpeting or burning a German ship or two, or some German barracks.

Denmark had not been singled out for special treatment by SOE. In June 1941 statements of the potential for subversion in all occupied European countries were called for, on the ground that special operations would form part of the general war effort and must conform to the overall strategic plan. All Country Sections were asked to prepare estimates of the stores, equipment, and transport they would need to establish a secret army and a sabotage organization in the country with which they were concerned. At the same time Turnbull, who had hinted in his first weeks that he had had less guidance than might be desirable, was swamped by a massive memorandum from London which left nothing to the imagination, and indeed appeared to assume that all it needed to establish a full-scale resistance organization in Denmark was the pressing of a few elementary buttons in the British Legation in Stockholm. The plan in Denmark must roughly follow the general plan

> to set up organizations in every enemy and enemy-occupied territory to harass the Hun in every possible way by the use of the following weapons: minor sabotage and passive resistance; major sabotage; armed mass revolt or insurrection; guerrilla warfare; raiding parties; propaganda and political subversion; and political assassination.

The proposed Danish manifestation of these weapons is then enlarged on. Although the Danes are already carrying out some passive resistance, it is sporadic. An organization for major sabotage must be formed, to lie dormant until required by the Allies. Existing resistance movements must be contacted, not only because they can make a contribution to the common cause, but to discourage them from taking premature action. Preparations must be made now for a mass revolt when the time is ripe. Although the Danish temperament is not suited to guerrilla warfare, it is a weapon that must not be overlooked. If raiding parties are sent in from Britain, guides must be ready to conduct them to their objectives. 'Our parachutist must know that, if he lands in a certain field, for instance, then the man who keeps the farm nearby is friendly and in touch with our organization.' The value of subversive propaganda must not be forgotten; and the minds of even pro-British elements in the community suitably cultivated. Political assassinations should be arranged so that they seem to be inspired and carried out by Danes. (In fact the Danish resistance set its face against political assassination.) One problem which this ambitious memorandum foresaw was the difficulty of stirring up the Danish people so long as they were well treated by the occupying power. 'The time may come when we shall have to employ political subversion to force the Germans to put in their own administration, in order that the Danes may feel the pinch and therefore become more willing tools in our hands.' While the memorandum does not specifically refer to a secret army, a basic idea in SO 2's thinking, it provided Turnbull with food for thought, and a programme which a superman would have found it hard to implement, even were he living in the country.

Turnbull responded by submitting his own scheme. Bodies of Danes would be enlisted throughout the country who would remain quiet for the time being, but would be ready to go into action at a moment's notice. He would divide his organization into six sections: to provide transport and guides for raiding parties; to carry out sabotage on the eve of a British landing in Denmark; to foster cold shouldering of the occupying forces; to establish communications with existing pro-Allied groups; to disseminate propaganda; and finally a section to dispose of Germans and traitors on The Day, in which each man will have his victim marked down and be responsible for liquidating him. This version of the 'excellent and helpful plan' provided by headquarters (Turnbull's assessment) was duly approved by London.

In July Turnbull saw a chance of establishing a secret army under the noses of the Germans. The amount of intelligence percolating from Denmark to Stockholm, meagre at first, was increasing thanks to the efforts of Ebbe Munck, the Danish explorer and journalist. Munck, who had been a foreign correspondent in Finland at the time of the German invasion of Denmark, went home to arrange with the Intelligence Division of the Danish General Staff for the supply of information about German military affairs, to which they had free access under the agreement with the occupying power. He then had himself posted to Stockholm, with the cover of correspondent for *Berlingske Tidende*,

Denmark's leading newspaper. He worked in close collaboration with the Stockholm Mission, providing a link with Denmark, including microfilm messages, which was invaluable in the days before SOE had its own lines of communication. Through this link Turnbull became aware that the Germans were considering forming a special Danish force which would enable them safely to reduce their own garrison; and he proposed that an SO team should be assigned to work among the Danish troops, and secretly divert their loyalty to the Allies. Nothing came of this ingenious scheme, and headquarters continued to think of a secret army run from London. It would be small, only 1,000 men for the whole country, divided equally among twenty-one districts. Munck then informed Turnbull of a Danish plan to mobilize a secret army from the ranks of the Danish regular forces. Stig Jensen (later chief reception organizer in Zealand), who was operating from the Jutland estate of Flemming Juncker (later chief organizer for Jutland), was in touch with the Danish garrisons, which had been left intact by the Germans as part of the fiction that Denmark was still a free country. Four garrisons 'are already with us [Turnbull meant with Stig Jensen] and are storing away arms and ammunition. Six other garrisons are shortly to be roped in.' SOE in Stockholm was training a W/T operator who would in due course be parachuted to Juncker's estate so that orders from London could be relayed to the garrisons. This would make it possible to co-ordinate the movements of the secret army with a British invasion. Arms and equipment would be sent in by air from Britain to supplement stocks already held.

At the same time it came to light that, even before the German invasion of 9 April 1940, some members of the Danish General Staff had discussed the formation of a clandestine force to operate against the Germans if they did invade. Munck told Turnbull about the plans of these men, known as 'the Princes'. In passing this on to London, Turnbull explained that they did not intend to build a large army but rather to organize groups of soldiers throughout the country so that, when the signal came for a general uprising, official Denmark as represented by the Services would be seen to take a hand. It seemed to him that this solved all SOE's secret army problems. He wrote enthusiastically to Hollingworth that the Princes were

> wonderfully placed for our work. They play in with the Germans and see and know all that is going on. These men are obviously in the best position to

organize our secret army, and I consider it a privilege to be the link between you and them . . . They are big men and therefore we cannot exactly give them orders!

Progress through Stig Jensen was likely to be slow and it would speed things up to deal with the Princes. Hollingworth replied that this was absolutely splendid and that he agreed with every word Turnbull had written. London still envisaged a force of 1,000 men spread over the whole country: twenty-one district commanders with four platoons of twelve men each. They calculated they would need 252 Sten guns, 1,000 automatic pistols, and 3,528 hand grenades, and proportionate quantities of other weapons and equipment. Turnbull must find out what the Princes could lay their hands on, so that deficiencies might be made good by SOE; and it was agreed that Stig Jensen's activities should be taken over by the Princes.

One of the Princes, Ritmeister (Cavalry-Major) Lunding, went to Stockholm for discussions with Turnbull at the beginning of March 1942; and again SOE's man reported with great enthusiasm about the contribution offered by the Princes. Lunding 'has brought a report of great progress and of something even more important and of greater scope than we had ever hoped'; but he had also explained that neither the Princes nor the Danish commander-in-chief, General Ebbe Gørtz, were prepared to accept SOE's plans for a secret army exactly as they stood. They offered instead a carefully thought-out scheme to mobilize picked units of the Danish army, which would be placed at SOE's disposal at the right moment. Since, under the constitution, the order to mobilize must come from the Minister of the Interior on the direction of the Cabinet—clearly impossible in existing circumstances—the commander-in-chief would take it upon himself to issue the necessary orders. One would bring between 300 and 600 men into each garrison, which they would have no difficulty in taking over as eleven were on a care-and-maintenance basis, and only Holbaek, Naestved, Randers, and Sonderborg fully manned. Letters to set this in motion were already prepared, and the men would be assembled within four hours of receiving them. A second order was aimed at members of specialist units who might live far from their reporting points; but even in their case half could be mobilized within four hours. Between them the two orders would make 8,000 men available almost instantly. Lunding claimed this meant that Denmark already had her secret army; and it would be nonsense to have two in a small country. They would almost certainly compromise each other. Turnbull agreed and accepted that SOE's army—code-name *Chair*—must now be abandoned.

Lunding did not stop at 'selling' the Princes' army. He used it eloquently as an argument for suspending all active resistance in Denmark. There must be no Allied raids on the Danish coast—indeed the Princes had already issued orders that raiders from Britain should be denied any help. He believed the Germans wanted to reduce their garrison in order to release men for the Russian front. This meant that, when The Day came, there would be in Denmark perhaps no more than one and a half second-rate divisions—an easy prey for the Princes' army when it came into the open. But any major sabotage would put their plans at risk, and therefore the dispatch of SOE agents into Denmark must stop. The most the Princes could agree to was occasional bombing attacks by the RAF, which would not lead to reprisals, since no one within the country could be held responsible. Their plans, which had the blessing of the commander-in-chief, were based on two possibilities: that the Reich would disintegrate from within; or that there would be a British landing in Denmark, when the army would collaborate to the fullest extent.

All this made sense to Turnbull, although it was questionable whether it was in the interest of the Allies—especially the Russians—to run down the German garrison in Denmark. He said he was delighted to know that SOE had the full co-operation of leading men in the Danish army; and agreed that parallel secret armies would be 'arduous and risky'. He burned his boats:

> At this juncture the Prince officially asked me to accept his chief's [Gørtz's] plan; and I felt I had headquarters' approval in saying that we were extremely glad to be able to co-operate with General Gørtz and that I felt sure that my headquarters would not mind altering their own local plans in this case to bring about full collaboration with General Gørtz and the Princes. On your behalf therefore, since time was short and the Prince especially asked me not to telegraph anything of this, I agreed to scrap the old *Chair* plan and to base all our future activities on collaboration between the Danish army—albeit in a special form—and ourselves.

The Princes' objective was to keep on lulling the Germans into a sense of false security so that, when finally The Day arrived, 'their plan will break on the Germans like a clap of thunder'.

Turnbull believed that, if this effective ready-made local force was to be preserved intact for future use, Britain would have to be very circumspect in her policy *vis-à-vis* Denmark. At all costs the *status quo* must be maintained. The Germans must be left with the impression that Denmark was a model satellite state. The King must remain on the throne, the government must continue in office, there must be no rival government in exile, political unrest must be avoided. Above all, there must be no large-scale sabotage. To foster the idea that Denmark was resigned to the presence of the occupying power, subversive propaganda put out by Britain within the country should take the line:

> It is impossible now to say whether in April 1940 Denmark could have opposed Germany as did the Norwegians. In any case it is now too late to follow Norway's example, and it is necessary to lie low. The present is a time of humiliation . . . We must take the consequences of what occurred two years ago.

As part of his passive programme, Turnbull recommended that the planned exfiltration of Christmas Møller, the country's leading anti-Nazi politician, to strengthen the overseas voice of Danish patriotism, should be abandoned. As Minister of Trade, Møller had gone along with the decision to accept the Germans, but he soon became an outspoken critic of the occupation, and lost his portfolio and seat in parliament: 'his stand had aroused particular attention throughout the country and also abroad. Everyone knew where he stood, and one could imagine what would be his aim.'[1] Turnbull thought his departure would cause a political upheaval that might end 'the democratic regime'; and so strongly did he feel the need to stop sabotage that he suggested that the dispatch to the field of Christian Rottbøll as chief organizer should also be put off.

London was equally carried away. There were great possibilities. Sporborg endorsed the file: 'I entirely approve.' Hollingworth agreed with the Princes on all points, although he did hint that they were unduly scared about the effect of a sabotage campaign on their army; but, since what was contemplated was nothing less than a secret military alliance, the whole matter must be examined at a very high level. His advice to that high level was:

> In the Princes' organization we now have an instrument of which one day we may be able to make vital use when the time is ripe to strike hard and swiftly at the heart of Germany. The arrival of Christmas Møller in this country would have an overstimulating effect on the Danish people and might be disastrous. I have always felt that the Danes were, of their own volition, moving in the right direction and at the right pace. Hitherto our plans for Denmark have been purely theoretical but now that we know for certain that there exists machinery

which can be put into operation within the space of minutes, it is important that we should reserve our efforts for one final blow and refrain from forcing the Danes to premature action.

Therefore, SOE should supply equipment to the Princes' men, divert to their secret army instructors now undergoing training, hold back major sabotage until the secret army went into action, and in the meantime allow only minor sabotage that could be made to look accidental.

In putting the proposal forward Sporborg said it must be accepted unreservedly, with only essential modifications. The real question was whether a secret Danish army was wanted badly enough to justify restricting activities in other spheres in Denmark, and he had no doubt it was. 'I do not think we shall ever do very much in Denmark except on secret army lines.' If the present Danish government was forced to give way to a full-blooded Nazi regime, the secret army would be invaluable. The Princes must be told to prepare for this eventuality. A colleague viewed the proposition with slight misgiving. 'Although they do not state what reasons they may have for believing in the bona fides of this proposal, Turnbull, Hollingworth and Sporborg clearly believe it to be genuine.' If they were right, it should be acted on, with all that that entailed. It would mean settling for little or no sabotage in Denmark, which was a pity, but it 'would be no worse a fate than we have had to bear in Czechoslovakia, and, until recently, Poland'. The Joint Planning Staff (JPS) who had recommended positive support for secret armies in France, Belgium, Holland, and Norway suggested there should be no more than moral support for the Princes, since it was unlikely that a British expedition would ever be sent to Denmark. Moreover, the level of sabotage should be kept up to discourage the Germans from withdrawing troops.

On 16 April 1942, shortly before he resigned on grounds of ill health, Sir Frank Nelson issued a directive on the basis of this rather contradictory advice. The Princes should be encouraged so long as it did not affect SO 2's work elsewhere in Europe. As to the Princes' request that British propaganda should be toned down, and sabotage activities ended or greatly reduced to encourage the Germans to reduce their garrison, any such reduction would be contrary to the wishes of the COS. Moreover, Denmark flanked one of the main Axis shipping routes and provided excellent opportunities for sabotage. The Scandinavian Section would therefore negotiate with the Princes for the reception of W/T equipment and stores to the extent they could be spared; and for the maximum scale of sabotage they would agree to, which could be carried out either by a separate organization under SOE control, or by the Princes' organization, or by part of their organization under the direction of SOE. Of the three alternatives he preferred the last, and the main targets should be shipping, train ferries, and all supplies proceeding to and from Germany.

When Hambro succeeded Nelson as CD a few days later, doubts were raised about the consistency of SOE's Denmark policy. On the one hand it appeared that the Danish Section contemplated only minor sabotage in Denmark, whereas CD had asked for the maximum sabotage the Princes would agree to.

> In view of the indication given in previous reports regarding the Princes, that they are concerned primarily with a plan for taking over control as the Germans retire from the country and that they are unlikely to indulge in any sabotage which might provoke the occupying forces to any form of retaliation, it is not at all clear how our sabotage organization in Denmark can have all its action co-ordinated with the operations of the secret army—unless our people are successful in persuading the Princes to change their policy.

The boot was on the other foot. The Princes were desperately anxious that SOE should change *their* policy. They were adamant that a sabotage campaign would lead to the destruction of their organization, and pleaded with Turnbull to make their case to London. They claimed, for example, that a proposed attempt to sabotage the Copenhagen–Malmö train ferry (Operation *Barholm*) would bring serious reprisals, but would accomplish nothing. They could easily do the job themselves, but the game wasn't worth the candle. More than twenty replacement vessels were available and traffic might be interrupted for a week or at best a fortnight. Turnbull asked that the dispatch of further agents should be held up, but London was unsympathetic. They would continue to send in agents, but they would attack only vital targets, the Princes being consulted in each case—a small concession.

Although the Princes guaranteed to put 8,000 men into the field at short notice, Lunding provided Turnbull with figures suggesting that in the longer term a far greater force would be available. The 20 age group liable for military service totalled about 140,000 men, of which about 30,000 were working in Germany. This left a balance of 113,665, a remarkably precise figure. When it was put to SOE's planners, they treated it seriously and came up with the assessment:

> The number of sorties required to provide the Princes' things [i.e. arms and

equipment for an army of the size specified] would be not less than 4,200, 6 containers being carried in each sortie. As it is only possible, under ideal conditions and assuming that the Princes could arrange a sufficient number of reception committees, to do 10 sorties a month, it would take 35 years to equip them. This astronomical figure is of course out of the question.

The best that could be done in the next nine months was to send arms for about 500, half the number of SOE's originally projected secret army.

In the first week of August 1942 Major Gyth, the number two Prince, came to Stockholm, when Turnbull met him secretly three times. The first and third meetings were *tête-à-tête*, but Colonel Larden, then head of SOE's Stockholm Mission, attended the second, and conveyed the good wishes of the British COS to the Princes and General Gørtz. Gyth painted an even more glowing picture of the strong position of the Danish army. The Princes could call on enough men at short notice to destroy or deny to the enemy all major points of communication, and to contain the German forces until the whole secret army was in the field. Men were living near every bridge and other important points, carrying on their normal work as railwaymen, clerks, postmen, etc., all instructed as to their exact task. Secret orders were already lying at every garrison; and at the four fully manned garrisons there were ample stocks of munitions at the disposal of the army. At Holbaek, for example, there were 8,000 rifles, 3,000–4,000 machine guns, 12 75 mm field guns, 4 long-range 10.5 cm guns, anti-aircraft guns, and tommy guns—all with the necessary ammunition. Lorries were standing by with full petrol tanks, ready to carry out their sealed orders. For example, one might receive an order to rush rifles and machine guns to a special farm near Frederikshavn. 'The whole thing will go ahead quickly and silently and between one and five hours it is guaranteed by the Princes that all points already chosen will be supplied with the requisite equipment, uniforms etc., probably without the Germans getting wind of it at all.'

On the other hand, the Princes continued to oppose anything but minor sabotage. They emphasized that conditions in Denmark were unique. No other occupied country still had its army intact, and it was most unfair to saddle Denmark with plans which might be appropriate elsewhere. It would be a tragic waste to put their secret army at risk for the sake of a hypothetical short-term advantage. At the second meeting Larden argued strongly in favour of the opposite view. Immediate sabotage of communications might have a more direct effect on the course of the war than a hypothetical *coup d'état* when the Germans were already well on the way to defeat. Gyth admitted there was some force in this argument, but repeated that any sabotage campaign must lead to 'a squandering of the unique resources which have been collected in their country for a knockout blow on the chosen day'. It was agreed to refer the matter back to London, pointing out that the advantages of major sabotage were short-term dislocation of traffic to Norway, and perhaps forcing Germany to send in more troops to Denmark; and the advantages of only minor sabotage were that the secret army would remain intact, and the flow of intelligence supplied by the Princes, a carrot of great importance to the Allies, would continue uninterrupted. Once again Turnbull passed on to London the Princes' impassioned 'hands off' plea.

Slowly—surprisingly slowly, it may be thought—scepticism about the Princes' secret army developed in SOE headquarters. Although both Turnbull and Larden were 'very favourably impressed with Gyth's sincerity and frankness', London began to have its doubts. One proposition which had been accepted without question in Stockholm was that the Princes were in touch with a gun runner who was able to supply them with 'anything from rifles and Tommy guns to field guns, and he can draw much of this equipment from Germany itself'. Gyth, however, refused to disclose the name of this man, or to allow Turnbull to get in direct touch with him. SOE did comment that this sounded like a fairy-tale, but even so gave the Princes the benefit of the doubt: 'Presumably the Prince concerned is no fool.' No one seems seriously to have asked why, if they could get all the weapons they needed from their own arsenals, they should have to rely on a gun runner, or even on supplies from SOE.

It was Hollingworth who first began to suspect that the Princes might be overstating the case. He confirmed that the intelligence they supplied was first rate (to be expected since it came straight from the machinery of the Danish General Staff), but went on to say: 'We are, however, taking too much for granted if we accept in blind faith the Princes' guarantee that their *coup d'état* will have the promised effect.' There was a report from the field that the Germans knew quite well what the Princes were up to, and that they would step in before the secret army went into action. Nevertheless, the Princes claimed 'that even if they should be hindered in their movements their organization would not fall to pieces. It was covering the whole of the country and was sufficiently strong.' Gubbins then took a hand. After analysing the situation in Denmark, he wrote a memorandum which concluded:

We are in touch with an organization which alleges that it can take over control of the country by mobilizing the Danish army at the time of an allied invasion of Denmark or when we inform them that a serious return to the Continent is being undertaken . . . The leaders of the above organization view with disfavour any acts of sabotage likely to compromise the existing situation, and state that under present conditions it can remain self-supporting in weapons required for those groups which we had previously agreed to supply.

This policy, said Gubbins firmly, does not satisfy me.

He ordered the Danish Section to spend the next six months developing its sabotage organization and communications in Denmark, which, whatever the Princes might say, were of the utmost importance. There would be three aircraft sorties a month to drop W/T operators, sabotage instructors, sabotage equipment, and propaganda material. The maximum sabotage would be carried out against all forms of transport, especially railway locomotives and rolling stock (but not railway lines), and establishments working for the Germans. SOE would try to fit in with the plans of the Danes' secret army, but it must insist on its right to carry on with sabotage.

In October it was decided to increase the pressure on the Princes. Turnbull was instructed to put them on the spot by telling them that London was 'especially keen' that something should be done in Denmark to help the war effort. He wrote to them:

we wish to point out to you, SKOV [the new code-name for the Princes], that you are the only party in Denmark who can now seriously damage the German transport system. You told us that, if something had to be done, you could do it much more efficiently and competently than *Table* [SOE's own sabotage organization]. We therefore appeal to you to make some positive contribution to the result of the war now or in the immediate future.

He asked them to recommend three targets which they could destroy without causing too much internal trouble—ferries, bridges, railway junctions, and locomotive works were obvious targets, but the choice would be left to the Princes—and, when the date and time of the operations had been settled, SOE would arrange for the RAF to mount a bombing raid against the same objectives so that the Germans would be in doubt as to the true cause of the damage. Turnbull rather weakened his plea by adding that, if Denmark never did anything, it would be the worse for her after the war. The Princes must already have carefully weighed the pros and cons of their refusal to undertake sabotage—indeed their positive objections to sabotage by any group—and have decided that it was better for Denmark to try to go through the war unscathed, if they could contrive it. They continued to claim that in the long run their army would accomplish much more than SOE sabotage could ever do.

London's scepticism about the motives of the Princes continued to grow. At the beginning of October Gubbins wrote to Larden:

> Myself, I still have to be convinced that SKOV has any real intentions to undertake active offensive work at any time. My reading of the position is that they wish to keep their country absolutely free of any sabotage or anti-German action until the Germans break up altogether and begin to leave the country and then in point of fact we shall not need them . . . SKOV is, I think, playing a very careful game of *via media* between us and the Germans. That is not good enough for us.

When, in reply, Larden said that the Princes had undertaken if necessary 'to come out in force' at the end of the year, Gubbins said there was no intention of calling out their organization. All he wanted was proof that SKOV was a serious resistance group 'of which I still have strong doubts. If SKOV can give definite proof by taking action that they are prepared to run risks, risk their lives, etc., at this moment, then we might be able to place a little bit more confidence in them.' At present Denmark was an asset to Germany. It was important that she should be made to become a liability.

In spite of the pressure, the Princes devoted the winter of 1942–3 to masterly inactivity so far as sabotage was concerned; and, when the leading Prince, Lieutenant-Colonel Einar Nordentoft, came to Stockholm in March 1943, he had to face a much more positive ultimatum. After expressing the Allies' gratitude for the Princes' continued supply of intelligence, Turnbull said that something more was now needed. SOE had kept sabotage to a minimum at the instigation of the Princes; but it had no doubt that sabotage now was worth much more than possible military action when the war was virtually won. So London were giving them six months to make a contribution. Nordentoft was shocked. Did the British government trust them so little as to insist on some act of sabotage, just to prove their good faith? The demand was mean and unreasonable. To suggest that that naughty little boy, Denmark, should atone for his sins in this way was petty and unworthy of the British character. He and his colleagues were Danes first and foremost, and must defend Denmark's interests at all times as much as, if not more than, the interests of the Allies. There would be no question of sacrificing their secret army for the sake of one factory here, or one goods train there. Surely the supply of intelligence was enough, especially as other organizations were carrying out acts of sabotage almost daily?

This was plain speaking, and Nordentoft's violent protest suggests anger at having his bluff called. Before he left Stockholm, he told Ebbe Munck that the Princes *had* been responsible for some minor sabotage, and, when a sceptical Munck asked why he had not told the British, he replied that it was on security grounds, hardly a convincing explanation. He also unburdened himself generally to Munck. The Princes had acted in good faith throughout, but, since Britain did not trust them, they might well decide to sever their connections with SOE. This petulant threat did not mean that they would take on the occupying forces single-handed but merely that they hoped to get through to the armistice without violent conflict in Denmark. London's reaction was gentle. They sympathized with the Princes' wish to keep their army ready 'for a wholesale and gratifying gesture on The Day, which will avenge once and for all the collapse of 1940'; but it did not fit in with Allied policy to damage Germany *now*. And in July, when SOE was still speaking of 'our' organization *Peter* (one of the alternative names for the Princes), the largest able to play a part against the Germans, it was concluded that it was reluctant on political grounds to encourage guerrilla activities. The Princes' policy 'has been diametrically opposed to that of the guerrilla groups with whom we have been in touch'.

It was felt in London that the time was ripe to increase pressure on the Germans; and by the end of July 1943 there were many successful major attacks. The JPS recommended they should continue, leading to 'an interim climax' in October 1943, followed by an all-out effort in the spring of 1944 on the eve of the Allied invasion. SOE felt it would be impossible to implement this plan with any precision. German reprisals for the 'interim climax' would leave the resistance movement so disheartened that it could not be rebuilt before it was really needed in the spring. The Foreign Office doubted whether increased sabotage would lead to a German take-over, but, if it did, many Danes would for the first time decide to take an active part in the resistance. There was general agreement that so far they had done nothing to justify even a back seat at the peace conference; and, if they wanted any sort of a seat, they should now show genuine support for the Allies. There must, therefore, be a 'gradual inflammation of Danish public opinion' that would come to a head a month or two before D-Day, when it would be essential to keep as many German troops as possible away from the Normandy beaches. This meant putting something of a brake on sabotage in the immediate future, which was deprecated by men in the field, although they loyally obeyed the order to lie low. 'This evidence of discipline is most encouraging . . .' At the beginning of August, however, it was believed that the 'inflammation' scheduled for just before D-Day was dangerously imminent, thanks to discontent among the people at large. In an attempt to save the situation, it was proposed that all sabotage should stop. Flemming Muus, who had parachuted into Denmark in March 1943 to act as SOE's chief organizer, had different ideas. He asked Nordentoft to co-operate in increasing pressure on the Germans; but the leading Prince said on the contrary he would try to calm things down. Earlier Muus had told London that, if the current rate of sabotage continued, the Germans would be forced to take over the country within three weeks; and, as it happened, factors outside SOE's control changed the whole picture.

The principal was the rivalry between Dr Werner Best, chief representative of the Reich in Denmark, and General von Hanneken, commander-in-chief of the German forces, a rivalry which had persisted ever since they had taken up office late in 1942. Best had been told by Ribbentrop to keep Denmark quiet, and he had therefore followed 'a policy of sweet reasonableness, which was in marked contrast to the somewhat brutal and truculent attitude taken up by Hanneken'. Despite the latter's attempts to undermine him, Best stuck to his gentle policy, but failed to keep Denmark quiet. As sabotage and general unrest increased, Best was ordered by Berlin to take tougher measures. Saboteurs must be tried by German courts in accordance with German law, and those convicted must serve their sentences in Hamburg. On 9 August, five days after they had received this ultimatum, the Danish Cabinet unanimously rejected it.

Meanwhile sabotage and civil disorder continued unabated. When the Germans posted anti-sabotage guards in factories working for them, it led to more strikes and disturbances, which in turn led to the declaration of a state of emergency in many parts of the country. The worst trouble was in Odense. A riot was sparked off when a group of drunken German soldiers shouted abuse at a Danish military band. An officer who loosed off his revolver in the scuffles that followed and hit a young girl was thrown to the ground and seriously injured. Later shops and places of business belonging to Danish Nazis and those with German sympathies were ransacked. Girls who had been friendly with German soldiers were stripped and painted with swastikas. There were similar disturbances in other provincial towns. The German garrison in Copenhagen was reinforced at short notice by the diversion of troops in transit to and from Norway.

On 28 August Best, who had been summoned to Berlin four days earlier for consultations, presented the Danish government with two more ultimatums. First, Odense must pay a fine of a million kroner, and those who had attacked the German officer (who had died) must be handed over. If they were not, ten citizens would be taken hostage. Although it was not spelt out, that meant they would be shot. Second, the Danish government must declare a state of emergency and prohibit strikes and public gatherings. There would be censorship of the press. After 1 September 1943 saboteurs and any caught with firearms or explosives would face the death penalty. These demands were rejected out of hand. The government resigned, and on 29 August, three weeks to the day after Muus's warning to London, the Germans took over the whole administration, prohibited parliament from sitting, and disbanded army and navy. The King and Queen were under house arrest in their palace. High ranking officers were required to report to the Germans by 3 September.

SOE allowed itself to think that this was the moment the Princes had been eagerly waiting for. At least one European country would be set ablaze in fulfilment of Winston Churchill's command of July 1940; but, although they had had ample warning of the Germans' intentions, the Princes did nothing. The orders supposed to bring a powerful force into the field in a matter of hours were never issued. Not one man was asked to lift a finger against the Germans. Nordentoft and Gyth went underground and later escaped to Sweden. The third Prince, Lunding, was arrested in his office and taken to Germany for questioning. The fourth, Major Per Winkel, gave himself up on 3 September. In fact the Germans did not know of his association with the Princes, and he was freed, to join the others in Sweden later.

All that was saved from the wreck was the remnant of the Princes' intelligence service, which had been of great value. It was reorganized by the 'reserve Princes', whose leader, Svend Truelsen, had been secretly appointed in April 1940 to assume command of the intelligence services should the armed forces be disbanded.

> Truelsen completely rebuilt the intelligence service within a few weeks, expanding its sources to include contacts in northern Germany, and speeding up both the gathering of material . . . and its interpretation and assessment. He had to leave Denmark in May 1944. He was summoned to London, promoted to major, and put in charge of the Danish intelligence service at SOE headquarters.[2]

Intelligence had always been the Princes' soft option, and it is interesting that, in looking forward to the day when the Germans would remove the velvet glove, it was only the soft option that they decided to perpetuate. There was no 'reserve Prince' to take command of a fighting organization and put it at the disposal of the Allies, if the commanders were arrested or otherwise fell by the wayside.

It might be expected that in the circumstances the Princes would now be more amenable; but no. It was agreed that Gyth should come to London from Sweden in October 1943 to discuss the incorporation of the remnant of the Princes' organization into SOE's groups; but his main objective turned out to be to raise money for a force of 300 Danish officers and men who had escaped to Sweden and were being groomed to deal with any Communist activity in Denmark when the Allied invasion took place. This proposition found little favour with the British authorities; and Gyth for his part was equally unimpressed with SOE's plans for Danish resistance. He wrote scathingly to Nordentoft:

> Unless everything I have learned in twenty-two years of military studies, and everything our officers have been taught at the *École Supérieure* and the *Kriegsakademi* in Berlin, and all the military papers, including British, are wrong, there can be no doubt that the war [in Denmark] will now be decided entirely on the military issue.

Yet he found the Danish Section interested in one thing, and one thing only—sabotage. He also wrote to Hollingworth:

> I must remind you that it is one thing to collect and organize saboteurs, consisting mostly of people whose emotions are stronger than their brains, and another thing to organize a country of four million inhabitants to a situation which can only be described as war.

Gyth had a cool reception from the Danish Section. Hollingworth asked himself whom the Princes now represented, and decided it would not be the comrades they had abandoned in Denmark, who would have little respect for them.

For two years they did everything possible to prevent us carrying out our policy of sabotage on the grounds that their plan for resistance which should be put into operation the moment the Germans took over the administration was too important to be jeopardized. When that day arrived on August 29th they made no attempt to carry out even part of their plan.

This was a little unfair. The ostensible purpose of the Princes' organization had been to step in either when the Germans withdrew, or the Allied invasion began, neither of which had happened. On the other hand, if their secret army had been all the Princes resolutely claimed, it could have been put in the field in a matter of hours to do enormous damage to the enemy, perhaps at very heavy cost—but wars are not won without casualties. The important consequence of Gyth's visit was to confirm SOE's growing suspicion that the Princes' real objective had not been to help the Allies at a critical stage of the war, but merely to hold down the Communists after liberation. Muus had no doubt about this; and it was confirmed by Gyth's statement that the Danish refugees in Sweden '. . . were to be earmarked for suppressing any communist activity in Denmark on the occasion of an allied invasion or voluntary German withdrawal'. This is discussed by Ulf Torell:

An attempt by Gyth to induce SOE to contribute to the financial solving of the plan [to suppress a communist rebellion] was a complete failure. At this time the British didn't want to contribute to Danish officers sitting in peace and quietness in Sweden in order to suppress the communists at the end of the war, while others were struggling against Germany for the liberation of Denmark.[3]

Further, although there were still officers left in Denmark who could in theory mobilize a force on the lines originally promised, there was now serious doubt whether it had ever been a practical proposition. Gyth told SOE it would be unwise to rely on the reservists remaining in Denmark. They had never been told they were members of a secret army in spite of the Princes' assurances that thousands were standing by, ready to blow up this factory or that goods train. Gyth's admission seemed to confirm that the Princes' secret army was no more than a blind to minimize SOE's activities in Denmark and enable the country to survive the war unscathed.

Christmas Møller had been called in to reason with Gyth, who had been instructed by Nordentoft to have nothing to do with him. It took two meetings for Gyth to see reason. Thereafter Møller wrote to him, so that there could be no misunderstanding:

> From the very first day I came to this country I have stated that what Great Britain wanted done on military grounds we would carry out unquestioningly as a military order. Instead of Great Britain we must now speak of allied high command, and what it demands done must be done in accordance with orders given. We may submit our advice, suggestions etc., but we Danes cannot have a position different from that of the real allied nations, who must all obey, great and small alike.

Gyth now co-operated, albeit reluctantly. He was induced to sign an undertaking, a curious international document in which the preamble was longer than the substance of the treaty:

> I, Major Gyth, as representative of the Princes in Sweden, and in the belief that the Princes represent the Danish General Staff, have read and agreed the following conditions which are deemed essential by the Supreme Allied Command as a basis for the consideration of any co-operation between the Supreme Allied Command and the Danish officers in Sweden and Denmark.

After this gobbledygook, the 'treaty' simply said Danish forces would co-operate with and obey the orders of the Supreme Allied Command. The transaction did not lead to a *rapprochement* between the two Danes—if anything the gulf widened. In a letter to Erling Foss, Gyth wrote that Møller was 'a very little man', the symbol of disintegration, who criticized everything and everybody.

However grudging the Princes' agreement to subordinate themselves to the Allied high command, it removed an important stumbling block to the Danish Section's activities. Hollingworth had feared that he would have men in Denmark supposedly working for him although their attitude to resistance was diametrically opposed to SOE's, and who claimed allegiance to an unrecognized and unreliable refugee group in Sweden. Turnbull, with the benefit of hindsight, thought it might have been wise to curtail sabotage even more, but accepted that the 'spontaneous outburst of popular feeling' would probably still have tipped the balance and forced the Germans to take over. He added that in any case Himmler, recently appointed Reichminister of the Interior, would almost certainly have insisted on a tougher line in Denmark, sabotage or no sabotage. The Stockholm Legation's considered verdict was that the Best/Hanneken contest which Best lost (mirroring a Himmler/Ribbentrop contest in Berlin) was the real key to the change in German policy which ended Denmark's comfortable ride and SOE's plan to regulate sabotage so that it reached a crescendo on the eve of D-Day.

Whatever the reasons, the consequences of 29 August 1943 were far-reaching; and they were irrevocably confirmed by a German act of political madness at the beginning of October. The introduction of martial law gave the German administration a completely free hand for the first time, and Best now planned to deport the whole Jewish community to concentration camps in Germany. A minor civil servant working for the Germans gave advance warning of the plan with the result that the great majority of the 7,000 Jews were with the help of the resistance and the whole Danish people enabled to escape to Sweden. But 472 who failed to slip through the German net were sent to the concentration camp at Theresienstadt in Bohemia. There were two important results of the Germans' error of judgement. The number of escape routes to Sweden was vastly increased, with consequent benefit to SOE's communications; and any remaining doubt in the mind of the Danish people about the wisdom of affording the maximum help and encouragement to the resistance was finally swept away.

7
Sabotage: Denmark

In June 1941 SO 2's Danish Section began to train agents and plan for activities in Denmark. The country was divided into six sectors each to have its own organizer, subdivided into twenty-one areas where a series of seven-man cells would operate. Two main organizations—*Table*, for sabotage, and *Chair*, the secret army (see Chapter 6)—were to be supported by related operations: *Dresser*, to establish lines of communication; *Settee*, to raise funds to finance the resistance groups; and *Divan*, to provide intelligence for sabotage activities. Finally, there was *Chest*, an organization to disseminate subversive propaganda. These enterprises were collectively known as *Booklet*.

After training, Dr Carl Johan Bruhn, chief organizer of both *Table* and *Chair*, would go in by parachute to act as chief organizer. Sabotage instructors would follow to recruit agents, reconnoitre dropping zones, organize committees to receive men and arms, and suggest targets for the RAF. There would be W/T links with the Stockholm Mission and London. Bruhn was an admirable choice. Aged 37, he was married to Dr Anne Connan and during the last weeks of his agent training, which he completed with flying colours, he also qualified as a doctor. The commandant at the Beaulieu training school thought he would achieve great things. 'A highly intelligent, very able man . . . should make a first-class organizer.'

At this time the Whitehall interdepartmental committee on Danish affairs was critical of the lack of progress in Denmark; but SO 2 defended itself by pointing out that the Germans were still treating the Danes very well. There was no hope of widespread subversion until they were made to feel the pinch. It might be possible to put the screws on the Danish government through an all-out propaganda campaign that would force the Nazis to take over the whole administration of the country; but the timing must be right. There was no point in producing a climate for sabotage if there were no trained men ready to take advantage of it. The attempt to force the Germans' hand should be timed for March 1942, by which date arms dumps must be prepared and trained agents ready to go into action.

Bruhn was to be sent in with a W/T operator, Mogens Hammer, at the end of October 1941 (Operation *Chilblain*). Elaborate arrangements were made to receive them. A small resistance group in Denmark was already in touch with Ronald Turnbull in Stockholm and would indicate a dropping point with two white lights and one red, set in a triangle 300–400 yards apart, with the red down wind. When the aircraft spotted them it would wink its navigation lights, approach up wind, and release men and their equipment containers. There would be four men on the ground to receive each container, and others to hide the parachutes quickly. Alfred Christensen, born in England of Danish parents, who was SO 2's man in Gothenburg in Sweden, provided addresses of safe houses in Denmark and a system of communication through 'family letters'. The BBC would warn the reception committee that *Chilblain* had been launched by broadcasting a talk in their Danish service ending 'You are either for or against Hitler'.

Except for the BBC code message, none of this was required. Bruhn elected to drop blind—without a reception committee—at Haslev in a wooded area he knew well. He had studied forestry there and had friends to whom he could look for help. He grew a beard, shaved his moustache, and planned to call at the house of one of his friends. He would say he was on the run from the Gestapo, and later would explain his mission and enlist his host's help. In the event the drop had to be postponed to the end of December because of bad weather.

Meantime other resistance movements were starting in Denmark. Professor Mogens Fog, a leading neurologist and 'extreme socialist', organized a subversive propaganda group which later blossomed into a full-scale resistance organization. Eigel Borch-Johansen, Secretary of the Small Shipowners' Association, had told Turnbull in Stockholm that he had founded a nationwide movement with the same aims as *Table*. He planned to attack coastal defences, occupy air bases, interrupt road and rail communications, and even incapacitate strong-point garrisons by introducing bacteria into their food. In passing this wildly ambitious programme to London, Turnbull assessed only the last item as 'a little fantastic'. Surprisingly, London agreed that Borch-Johansen should have a place in the scheme of things. His plans, although confused, seemed to be good. It was thought he would make a good deputy to Bruhn. He should be assigned to either *Table* or *Chair*, certainly not both. The secret army must not be put at risk by association with the *Table* sabotage groups.

However, SOE's plans for Denmark came to an abrupt halt when on 28 December 1941 Bruhn's parachute failed to open, thanks to a million to one chance which the parachute experts were aware of, but had discounted. The snap-hook anchoring the rope that opened the parachute was whipped off its anchor point before the rope did its job. (As a result of this accident the snap-hook was redesigned so that it could be locked on to the anchor point.) The man who found Bruhn's body in the snow guessed he might have had companions and therefore delayed telling the police, who delayed telling the Germans. This gave Hammer, who had landed safely, time to get away; but it was Bruhn that SOE had been relying on. Hammer was a W/T operator, not a leader. Nevertheless, London instructed him to do what he could to pave the way for Bruhn's replacement. Bruhn's death was notified to Turnbull in Stockholm in a pre-arranged code message in a Danish newspaper offering a reward for 'a green lady's handbag with clasp missing'. This later became a common method of communication.

The Princes, ever anxious to monitor SOE activities in their endeavour to keep sabotage to the minimum, complained that they had not been brought in. They admitted that they could not have prevented Bruhn's death, but at least they could have removed traces of the accident and kept Hammer's arrival secret. It was almost as if fate had played into their hands, for if Bruhn had arrived secretly it would have been some time before even their efficient intelligence picked up the fact. They said: 'If every detail is not thought exactly out . . . then something always goes wrong.' They wanted nothing to do with this sort of mismanagement. Turnbull was impressed. He feared it would antagonize the Princes 'if we press on wildly' and sent in more agents. London was less impressed. They sent in three more agents in February: Gunnar Christiansen (*Table Mat*); Adolf Larsen (*Table Tennis*); and Ole Geisler (*Table Manners*). (*Table* agents were given code-names associated with the word 'Table', until they ran out. No doubt an amusing ploy, but one that would help rather than hinder enemy counter-intelligence.) They were followed in April by three more: Captain Christian Rottbøll (*Table Top*), the 25-year-old nephew of the Danish Consul-General in London, who had fought in the Russo-Finnish war; Poul Johannessen (*Bates*); and Max Mikkelsen (*Cotton*). Since the three were unknown to their reception committee, they were to announce themselves respectively as 'Charles', 'Richard', and 'Fyrtarn', to which the response would be 'Chaplin', 'Tauber', and 'Bivogn'. ('Fyrtarn' (Lighthouse) and 'Bivogn' (Trailer) were the two leading Danish stand-up comics of the day, one tall and thin, the other short and fat.) Whether or not the appropriate greetings were used, the arrival was less than perfect. The trio landed in a wood, receiving minor injuries, and had to abandon their parachutes and W/T sets, which the police found. They made their way to Copenhagen and soon afterwards were in touch with London through Hammer, who had constructed his own W/T set to replace the one smashed when he landed.

The difficulties that had beset the Danish Section were now aggravated by differences in the field. Hammer claimed that one of the party that had received Rottbøll had been hopeless—'extremely nervous and frightened'—and must never be used again. The Danish Section was unconvinced and reprimanded Hammer for attacking a colleague in a telegram that had a wide circulation. Hammer himself came in for criticism. Borch-Johansen wrote:

> He is rather full of himself and considers that he is rather more than merely radio man . . . I guess he feels himself fit for being a leader of large plans and exciting enterprises, which he is discussing with a group of young craftsmen, editors of a small periodical . . . This is a dangerous company to consort with, and I should like to have him isolated. But this I think would bore him to extinction being cut off from his admiring congregation.

Borch-Johansen had no doubt it was only a matter of time before the police caught him; and he asked London to remind him about his duties.

In an effort to get SOE's men to work together, Hollingworth sent them personal letters. Hammer's was difficult since he had to remind 'his dear old friend' that he was only a W/T operator—although of course the job was important and dangerous. He must stop dealing with people who had not been vetted, which endangered his own life and the lives of others. He must accept Borch-Johansen, 'a splendid man to whom we owe a big debt of gratitude' as chief organizer; and it was hoped there would now be a spirit of mutual co-operation. To Borch-Johansen he said SOE had complete confidence in him. He hoped there would be no more trouble with Hammer 'but if a man becomes a real danger . . . he should be liquidated'. Mikkelsen and Johannessen were also assured of London's confidence; and Rottbøll was told he was headquarters' chief representative, responsible for the good conduct of men sent in from Britain. Borch-Johansen would consult him about all operations. Hollingworth confessed he should have told Rottbøll before he left England that he would share responsibility with Borch-Johansen. Turnbull was more worried about Hammer's ambitions: ' we must face the fact that certain of the more ebullient types chosen for our work are in danger of developing into Little Napoleons or Little Caesars if they are given too much rope.' The importance of teamwork must be drummed into future trainees.

Almost before the letters were delivered, *Table* suffered a new blow. Borch-Johansen was arrested for complicity in sending out Christmas Møller. When the Germans arrived, Møller had been Minister of Trade, but, because of his steadfast opposition to the Nazis, he was forced to give up office. In March 1941 his name was included in a short list of prominent Danes who might be brought to Britain to represent the Danish people, if not the Danish government; and now, in spite of Turnbull's misgivings that his departure would cause a political upheaval and alienate the Princes, he was smuggled under a cargo of chalk to Sweden *en route* for Britain, accompanied by his wife and young son. There was rejoicing in Denmark when it was known that this great patriot, who had helped to receive the first agents sent in by SOE, was safe in London and able to speak freely for Free Denmark. 'People's faces lit up in a smile everywhere where Christmas's name was mentioned . . .'[1]

This episode had one unfortunate consequence for the resistance. One of the detectives investigating Møller's illegal departure was Roland Olsen (*Richard*), who was co-operating with the resistance. He released Borch-Johansen on parole to attend to some personal business, but Borch-Johansen broke parole, and went into hiding, eventually escaping to Sweden. His loss was particularly regrettable since it was due to purely political activity, which Turnbull had consistently argued must never prejudice resistance operations. It was compensated for by closer contact with Mogens Fog, who had provided cars for the reception of Operation *Table Top*, and had given its members sanctuary in his hospital. He was regarded by Turnbull as the ideal choice to succeed Borch-Johansen. Møller, now safely in London, agreed.

SOE's next mission was *Table Talk*: three sabotage instructors, Erik Pedersen (*Bright*); Hans Hansen (*Crisp*); and Anders Nielsen (*Alistair*) were briefed to recruit and train saboteurs. They took in W/T sets to replace those lost by the earlier arrivals. Turnbull again suggested holding the mission up, out of deference to the Princes' known objection to sabotage; but London again refused. The three dropped on 31 July 1942 near the Overgaard estate of Flemming Juncker (a wealthy land-owner and industrialist, another early resistance leader) on the southern shore of Mariager Fjord. The drop was less than perfect, half a mile from a reception committee led by Rottbøll. Pedersen sprained an ankle and in the excitement of landing Hansen failed to recover his parachute. It was reported to the local magistrate who delayed informing the police. When the police arrived, they found containers with two of the team's W/T sets which the reception committee had failed to locate. The RAF was blamed for dropping them in the wrong place—the dispatcher had a bottle of rum, and was tipsy—but he was later absolved.

The reception committee took the parachutists' flying gear away by car; and provided cycles to take them to a safe house 30 miles away. Nielsen and Hansen later went to Copenhagen, where they found nobody at the safe house there and had an anxious time standing conspicuously 'with a wireless set in a big suitcase'. This was a bad start and it may have affected Nielsen's attitude to the organization: 'Everything was in an awful state. The leader (Rottbøll) spent most of his time arguing with the Princes. He was never in time for his appointments with us, always half an hour late if he showed up at all.' Rottbøll took Pedersen to Copenhagen after his sprained ankle had been treated. The train was searched by the police, who were aware that parachutists had dropped in the neighbourhood, but the fake identity cards they produced 'with calm confidence' were accepted as genuine.

Nielsen later confirmed that, while Rottbøll was leader, they accomplished absolutely nothing; but there were now signs that the organization was beginning to find its feet, as much due to the initiative of Danes within Denmark as to the efforts of SOE in London. A committee of three—Rottbøll, Stig Jensen, and Ole Killerick—had formed itself to run *Table*. The main problem continued to be Hammer, who refused to allow Mikkelsen to operate their only W/T set. Turnbull tried to defend him by referring to the special 'psychological factor in this type of work which gives the boys a sense of isolation and unreality and leads them to do things they would never think of doing under normal circumstances'; and Hollingworth appealed to Hammer, much as a tolerant father appeals to sons quarrelling over a football. He must show a 'sporting example' and allow the others to use the W/T. But Hammer was becoming a risk for other reasons. He had been recognized by an old school friend who had become a Nazi; and, in spite of the warning that he must stick to his own job, he had begun to collect intelligence. Rottbøll said the sooner he was out of the country the better.

The Germans had found out his transmission times and were actively trying to locate his station. By the middle of July they had pinpointed it to an apartment block, which they planned to raid hoping to catch him red-handed. But the Princes had heard of their plans and, when the building was surrounded, the bird had flown. Hammer's colleagues urged the obvious on headquarters: he must stop transmitting at regular times. The Princes made much of another breach of security. Mikkelsen and Johannessen's W/T sets, which the police had recovered, were labelled with their code-names—so Mikkelsen became *Tucker* and Johannessen *Toby*. Furious at this unprofessional performance, the Princes protested to London. All the *Table* agents had been dangerously indiscreet. Borch-Johansen couldn't keep his mouth shut. He must be got out of the country. They were now unwilling to have anything to do with men sent in by SOE.

Table's misfortunes continued when on 5 September Johannessen was surprised by the Danish police while transmitting. He had told the home station that he would send 100 groups, but after 60 signalled 'Wait' and there was no more. Before being overpowered he killed one policeman and injured four others. On the way to the police station he swallowed his suicide pill, murmuring 'Forgive me!' which Turnbull took to be a message to his colleagues. Once again the Princes said, 'I told you so!' They had warned Johannessen that the area where he was operating was being monitored by the Germans, but he had ignored the warning.

Both Borch-Johansen, who had succeeded in evading capture, and Hammer were brought to England. To Turnbull the former was no longer 'an immense asset' but rather 'something of an eccentric who indulges in flights of fancy'. Rottbøll accused him of letting the side down by promising much and achieving little. Nor had Hammer been a success. Like Borch-Johansen he 'had gone outside his mandate'. The efficient agent must think only of the job assigned him—'boastings and flights of fancy and impatience and self-advertisement have no place in this work at all'.

The next victim of *Table*'s atrocious security was Rottbøll himself. The police had found in Johannessen's flat a card index with the names of all the *Table* agents sent in from England which quickly led them to Rottbøll. When he was challenged, the revolver he was holding behind his back went off accidentally, and the police immediately opened fire, killing him instantly. This new disaster reduced SOE's organization to almost nothing; but the leading 'local people' were 'all standing firm, and I think are likely to ride the storm'.

The self-destruction in which the *Table* agents had indulged had to be stopped. In October 1942 SOE planned a new organization with much stricter ground rules. All association with earlier contacts must cease. There would be close collaboration with Mogens Fog, who would be entrusted with all *Table* activities in the immediate future. Later on *Table Top* and *Table Orange* (the code-name for the Fog group) would work together to produce the maximum effect and provide mutually helpful diversions. The crucial importance of security was stressed. The leader of *Table Top* would deal only with his deputy and Fog, in each case through two cut-outs. The two *Table* organizations would communicate with the Princes only through London or Stockholm, and all three organizations would maintain separate links with Stockholm. These new arrangements should be fully explained to Fog—nothing was to be held back; but the Princes should be told no more than was strictly necessary.

The most surprising feature of the reorganized *Table* was the appointment of Hammer as chief organizer, simply because there was nobody else. The Danish Section must have had serious misgivings, especially as Turnbull, not knowing that Hammer was the new organizer, continued to report unfavourably on him. He had revealed that he was a member of 'the parachute gang'. 'The fearful stupidity of Hammer and the others in mixing themselves up in *De Frie Danske* [a clandestine newspaper] is almost unforgiveable.' Three weeks after Hammer's return to Denmark on 18 October 1942, Turnbull wrote: 'Don't you think it is a pity for *Top* [i.e. the new organizer] to concern himself with propaganda? I remember only too well that it was tinkering about with secret newspapers that wrecked Hammer.' Some believed Hammer's 'talkativeness and boastfulness' were responsible for Rottbøll's death. Turnbull added: 'You will see that the new *Table Top* himself in his latest messages to you recommends that Hammer's brother [who had been guilty of indiscretions] should be spanked if he reaches the United Kingdom.' Stig Jensen was worried lest Hammer [presumed to be in England] should influence headquarters. He would paint a picture favourable to himself, and spin a plausible yarn. 'I am sure, however, you will not be taken in by any tall story.' This barrage of unwitting criticism of the new leader must have given Hollingworth sleepless nights.

The one thing nobody questioned was Hammer's courage. During his parachute training he had broken a leg and injured his pelvis, thanks to his considerable weight, so on his return to Denmark he chose to drop into the sea. This was safer from the point of view of weight, but made his reception more problematical. He arrived off shore in north Zealand, not 'the well-dressed commercial traveller' who had left Denmark but 'a round-shouldered bespectacled schoolmaster'. He reported his arrival in a telegram worthy of Alfred Jingle: 'Arrived according to plan. Operation great success. Dinghy not necessary. Did not use mine. Police have no suspicion about my arrival. *Orange* [i.e. Mogens Fog] eager to co-operate. *Peel* [i.e. Ole Killerick] the same . . . '

Hammer immediately sent Hollingworth a copy of *De Frie Danske*, thereby revealing he had not mended his ways. Hollingworth replied: 'For the sake of the more important work you have been entrusted to organize, please leave this work to others!' There already was a propaganda organization, and

> we must be absolutely sure that you are in no way involved . . . please stick to *Table* matters only . . . Please dissociate yourself from *De Frie Danske* and everybody connected with it . . . We have just received special information which makes it clear that your association with *De Frie Danske* constitutes an extreme danger to yourself . . . Please accept this warning: you are too valuable to lose!

It was unlikely that Hammer could hide his identity for long. Turnbull soon guessed that the new organizer was none other than the man he was blackballing. He told London he was visiting his old night club haunts, recognized by many. 'This is of course due to the crazy desire on the part of many people to make themselves interesting . . . his gaudy past is threatening here and there to catch up with him, and indiscretions committed previously are coming home to roost.' On one occasion 'he built himself up as something of a hero to a girl and told the whole story of his own experiences and those of his pals'. On another, when asked by *Richard* to return a key document he had purloined, Hammer said he had burned it, which was untrue. The document was later found in a raid on his flat and it was a miracle that *Richard*, whose help was invaluable, was not found out. In April the Princes said their piece. Hammer was a danger to everybody and must be recalled. If he refused to go, he must be flattered by the offer of an important new job in England. Peter Tennant joined the chorus. The men in the field thought Hammer should be liquidated. If headquarters could make such a disastrous appointment, they were capable of anything. Tennant added a comment on Borch-Johansen's organization. It existed only in his mind. Meantime three more *Table* agents had been removed from the scene. Pedersen, Hansen, and Mikkelsen were compromised, and Hammer decided they must be sent out; but they were arrested *en route* for Sweden and sent to Germany.

* * *

Happily there was now a ready-made replacement for Hammer. Flemming Bruun Muus had come to London from Liberia hoping to volunteer for a Free Danish Army. Instead he joined The Buffs and found his way to SOE. After training for nine months—much longer than his predecessors—he parachuted into Denmark in March 1943 with three sabotage instructors: Einar Balling (*Dance*); Vilfred Johansen (*Conduct*); and Jens Jensen (*Chatter*). Although they missed their dropping point, they were safely collected. At last there was a chief organizer with all the necessary qualities—tough, full of courage, ruthless, and above all security-minded. The first phase of SOE operations in Denmark had mercifully ended.

To begin with, however, Muus had to tidy up the mess left by Hammer, with whom he overlapped briefly, and who did not impress him. German penetration of the organization was deeper than anyone realized. All messages between London and Copenhagen in the summer and autumn of 1942 had been read. The Germans knew *Table* was a sabotage organization, and the identities of most members. They had found Johannessen's code, and they had learned a great deal from the interrogation of Mikkelsen, Pedersen, and Hans Hansen. Some *Table* agents continued to offer themselves as willing sacrifices. Adolf Larsen (*Table Tennis*), who had parachuted in on 13 February 1943, acquired a girl-friend in Frederikshavn, who was a notorious informer, and when she saw his revolver had no doubt he was not the insurance agent he pretended to be. She reported him to the police, who were helped by his adoption of the identity of a man still living in Frederikshavn. He disclosed the identity of two couriers, who were also arrested, one carrying a letter with a complete list of *Table*'s code-names. One of *Table*'s contacts in the police offered to smuggle a suicide pill to *Tennis*, but the offer was not taken up as it was feared he would give the pill to the Gestapo as proof that the police were working for the resistance. *Tennis* spent the rest of the war in German prisons, ending up in Dachau, where he was liberated by the Americans.

Another Larsen, Hans Henrik, was less fortunate. After training in Britain (when he was assessed by his instructors as 'intelligent, extremely keen and interested'; 'plenty of guts and initiative'; 'disappointing'; 'pleasant but rather colourless'; 'straightforward and reliable'; 'spoilt and very childish') he was dropped into Denmark and immediately began to drink heavily, and like Hammer to boast about his activities. To get rid of him he was ordered to escort to Sweden a girl wanted by the police, but he refused; and Muus, with London's approval, decided to liquidate him.

Turnbull reported the 'rather gruesome story'. Larsen was invited to a party and

> came in as usual in an inebriated condition and was very easily persuaded to accept a vermouth and water in which one L tablet had been dissolved. *Trick* tossed the drink off gaily and asked for more. The boys decided to dissolve two more tablets in a similar drink. Once again *Trick* gulped it down and his reaction was unexpected. He said he was feeling very gay and wanted to go out and dance. Once more the boys prepared another drink and *Trick* again drank the potion without ill effect. After the third dose *Trick* complained of sleepiness and went home. The next day he complained of a slight headache. Those present at the party say they were forcibly reminded of the end of Rasputin when his enemies had plied him constantly with various deadly poisons without any effect, until in the end they had to empty their guns into him.

Trick suffered the same fate. He was taken to the country, ostensibly to look for a dropping point, and shot.

Table complained bitterly to headquarters about the bland nature of the pills—or did vermouth act as an antidote? London replied with a rocket. They had failed to distinguish between the lethal and knock-out pills. Worse, if they thought they *had* dispensed the lethal pill for this purpose, they were all guilty of a serious breach of discipline by sacrificing their means of self-destruction.

The behaviour of the Larsens led Turnbull to philosophize about the selection of agents. *Tennis* had been tough, but not tough enough morally for 'this game'. His weakness might have been exposed had he been given £50 to spend on his 'dress rehearsal' in Britain. It might be better to choose well-educated men whose reactions under stress would be more predictable. The performance of men sent in SO 2 did not match that of the spontaneous resistance inside Denmark. Unless something was done about this there might be trouble.

> Above all, we must not let our good friends inside the country, who are many of them responsible persons with great experience in good positions . . . get away with the idea that our *Table* boys do nothing but harm to all the careful planning and painstaking security which has been undertaken by men like Flemming Juncker . . .

Certainly the profit and loss account for *Table*'s first two years made depressing reading. Of the ten agents sent in from Britain before Muus, four had been killed, four captured, and Hammer had returned to

Britain. The sole survivor of the original team—Anders Nielsen—admitted he had accomplished little, partly thanks to the shortage of explosives; and his service ended through a breach of security when he took a girl-friend with him on a tour of his contacts in Jutland.

It is remarkable how the leadership of one man changed the *Table* organization. Duus Hansen gave 'a glowing report' on Muus and even the Princes spoke highly of him. He had formed 'the boys' into a compact orderly team, which he kept well disciplined. One colleague said his presence had worked wonders. 'The old inefficient careless days of *Top* have gone and there is some order in things now.' As soon as Muus was firmly in the saddle, the whole tempo of special operations changed. Stores were regularly dropped and successfully retrieved. Sabotage proliferated. For the first time *Table* thought it was getting somewhere.

Sabotage carried out under *Table*'s direction included the destruction at the third attempt of the transformer at the Skandia railway carriage works at Randers, which had just received an urgent order from the Germans for 300 coaches, while diversionary incendiary attacks were carried out on neighbouring factories. A factory at Aalborg providing cement for German fortifications throughout Denmark was also put out of action by destroying the transformer, and machines used for bagging the cement were blown up. The shipyard pumping station in Aalborg was destroyed and the wall of the dry dock ripped open. Although the dock itself was out of commission for only a fortnight, it subsequently took up to three days to fill and empty it, compared with the usual three hours. There were successful attacks in Odense; on a rubber factory and an ironware works in Copenhagen; and on a joinery works producing solely for the Germans; and a German merchant ship was damaged by a limpet mine.

Not all the increased sabotage can be attributed to the arrival of Muus—some Communists were playing their independent part—but there is no doubt that Muus put new heart into the *Table* organization and into the 'resident' sabotage groups which were increasingly supplied with arms and equipment through *Table*. The Germans themselves had indirectly helped the resistance movement by allowing the Danes to hold a General Election in March 1943 after a relatively peaceful period, in the hope that the people were beginning to accept Nazi rule and that the old political parties would lose ground. The opposite happened. It was the pro-Nazi Peasant Party and the Danish

Nazis who lost ground and the old parties, led by the Social Democrats and Conservatives, were returned with a comfortable majority. All this suggested to the resistance leaders that their cause was far from hopeless. The way was now open for a strong leader to intensify the attacks on the occupying forces, knowing that more people than ever were behind him; and Muus proved to be just that strong leader. 'A ruthless and clever man, he made full use of his persuasive manner, a liking for bluff, and a shrewd assessment of the Section in London HQ, to flatter, bribe, cajole and drive the Danes to greater activity.'

In April 1943 *Table*'s own men were still involved in only minor incidents. They set fire to two wagon loads of straw destined for the German forces in Aarhus; and in Copenhagen they destroyed an oil storage tank and damaged a ship under construction for the Germans. The 'resident' saboteurs were more active. They seriously damaged a Copenhagen transformer station, destroyed by fire a flax-scutching factory along with its raw material stock, and attacked several industrial concerns working for the Germans.

In May *Table* damaged gas and water pipes at a factory manufacturing U-boats accumulators, a pumping station in Aalborg Harbour, and two German ships being repaired at Aarhus. Again the local groups working indirectly under *Table* had a bigger score. They successfully attacked an engineering works, a hall used for testing barrage balloons, shipyards where boats recently acquired by the Germans were burned, and a railway wagon works. The *Danske Industri Syndikat*, which was making machine guns, was seriously damaged. A carefully planned attack on a factory in Odense set the pattern for many later elaborate operations. The target was a four-storey building where electric motor parts were manufactured. The attack, planned and executed with military precision, was carried out on the night of 27 August 1943 by eleven men. The first obstacle was the heavy wire mesh protecting a basement window, which was cut through in three minutes while a cover party ensured that no stray passer-by came on the scene. The leader and a companion, each with 10 lb. of plastic explosive wrapped round his waist, waited inside the open window, while three others sauntered past at five-minute intervals, each tossing in another 10 lb. of explosive to be fielded by the men inside. They piled the explosive in a pyramid to be detonated by time pencils. Home-made incendiary devices were carefully placed some distance away to avoid damage from the main explosive. The time pencils were pressed, giving a safety margin of thirty minutes (if the timing was right!). If the night shift was warned too soon to evacuate the building, the Germans might have time to locate the charge and render it harmless, so the team waited as long as they dared—seven minutes—before telling the guards on the upper floors that they had ten minutes to clear the building. In fact they should have had twenty-three minutes, but the time pencils functioned prematurely, leaving only thirteen minutes for the evacuation of the work-force, who all got safely out. The building was completely destroyed, and burned for two days. It was rebuilt, and was attacked again next year when seven men simply drove up to the front door, and held the guard prisoner while planting their explosives. A large part of the new building was destroyed, again without loss of life, as there was now no night shift.

Reception committees were becoming more efficient, and bad luck (as distinct from bad security) which dogged the earlier *Table* seemed to have disappeared. Although Muus and his three colleagues had landed off target in March 1943, they were safely collected. Vilfred Petersen (*Mustard*), Jakob Jensen (*Pudding*), a sabotage instructor destined for Jutland, and Hans Johansen (*Brawn*), 'to do odd jobs all over the place', all landed perfectly. There was excitement when their reception committee had to hide while German troops fought a mock battle in the field where the parachutists were due to land. 'The Huns were exercising in our field all afternoon up to 20 minutes before the aircraft was expected. When they had "won" the exercise they, the Huns, cleared out, and we entered the field.' Poul Hansen (*Jelly*) who went to Elsinore, and Jens Carlsen (*Lard*) assigned to Odense, as sabotage instructors, 'landed nicely and everything went well'. Muus told Hollingworth that he made a point of not meeting 'the boys', but kept in touch through his cut-out Preben Lok Linblad (*Habit*). He admitted agents would prefer to deal with him direct, but the less they saw of each other the better. Hollingworth thankfully noted the file: 'For the first time we have a Field Organizer who rigidly adheres to the principles laid down by London.'

However, not even Muus was able to ensure perfect security. One of the men who dropped in with him, Einar Balling, a seaman trained as a sabotage instructor (*Dance*), worked first in Esbjerg, and later in Aalborg, where he shared a flat with Jens Jensen (*Chatter*) and Roland Lund (*Lamp*). On 26 August he returned to the flat to find six Danish police lying in wait. The only incriminating evidence they could find was his suicide pill, which he said he had found and had no idea what it was. He was sentenced to a month's imprisonment for failing to provide a satisfactory account of himself, and was then taken over by the Gestapo who were clearly well-informed about the *Table* men's code-names. They asked Balling if he knew *Chatter* and *Jam* (i.e. Muus); and was he himself *Lillefar* (Hollingworth's code-name at this time)? Balling 'acted as a stupid seaman' saying he had no idea what they were talking about, and was returned to prison. Next day he was rescued by four armed men and eventually escaped to Sweden. In the inquest into Balling's arrest it came to light that *Chatter* had been drinking heavily, that he had left papers with *Table* code-names in his flat where they were found by the police, that against standing orders he had taken part in both reception and sabotage, and that he had been sharing a flat with two other agents—all fundamental breaches of the security rules. *Chatter* had been present at a recent reception which was surprised by the Germans. One man was killed and another taken prisoner and it seemed likely that he had revealed the identities of all the resistance men present.

The ease with which Balling was got out of prison, and Muus's confidence that he would be got out, suggests that the escape was arranged with the connivance of the police to whom SOE was greatly indebted at this time. The second in command of the headquarters branch dealing with the underground movement, *Richard*, was one of the most helpful; but he could not have played the part he did had it not been that his chief who knew what he was up to 'looked the other way'. Three other members of the force greatly helped the resistance, all being put at serious risk by the criminal carelessness of some of the agents they were trying to help. Turnbull went out of his way to send *Richard* a message of thanks which so pleased him that it was said that from now on 'he would do absolutely anything for the *Table* boys'. In reporting this to London, Turnbull said:

> I suppose at times it must have seemed a very unequal fight to *Richard* and company but now I believe at last they are beginning to see results. Duus Hansen told me that he and *Richard* often discussed among themselves Denmark's reputation in England, and *Richard* said: 'You know, people in England must think we Danes are an awful lot of old women.'

On the other hand the police force contained people like the Andersen brothers, who were not brothers and whose only common bond was that they were violently pro-Nazi, and totally against the resistance movement. 'They are given most of the fruity jobs but *Richard* keeps an eye on them and there is at least one *Richard* man on duty with them, so he gets early warning of trouble.'

So successful had the dispatch of agents become that the supply began to exceed the demand. When Muus was told that four were being sent in during May he said it was too many. Until he got his organization into perfect shape, every newcomer constituted an additional danger; and he asked that only two should come in May. Muus wanted to build slowly in the interests of security. 'We are now well on the way to get the full co-operation of the communists and other groups and with the staff now here plus Jensen and Carlsen we are very well staffed to take care of all instructions for the time being.' Muus also allowed his dislike of propaganda to appear in this his first message from the field: 'Please exclude leaflets from the containers. It is not in our line to distribute them, while they on the other hand may constitute a danger to our security.' He also commanded: 'Please do not send chocolate, tea, coffee, cigarettes, etc. in the containers any more. It is awfully nice of you, but it may give rise to discontentment when one thinks he should have a greater share than the others.' A remarkable self-denying ordinance.

Turnbull reviewed the situation in Denmark on 2 August. Everything was well under control. Since the beginning of the year thirteen sabotage instructors had been sent into the field from Britain, of whom one had been taken prisoner and one liquidated for careless talk. The sabotage begun with the arrival of Muus had seen successful attacks on shipping and targets of importance to the U-boat campaign. Darkrooms had been set up in Stockholm and Copenhagen to handle microfilm communications traffic, which was now running at 100 messages a month each way. The organization for procuring currency was running smoothly. Many operatives were being recruited and trained locally. The four sections of the sabotage organization were working well: transport, including reception committees; security; billeting; and the evacuation of men for training in Britain and compromised agents. As soon as agents were received in Denmark, they were sent off to a safe house and had no further contact with the reception committee. Security was responsible for deciding when agents had been compromised, and for investigating what had gone wrong. W/T operators were protected from the German direction finding service by warnings from a friendly contact within the service. Although there had been daily exchanges with the home station, the Germans were under the impression that there had been no transmission for eight weeks. The coast-watching system set up for SIS was operating efficiently. As to the future, several new W/T stations would shortly be opened. There would be S-phone communication between Sweden and Denmark, lest the ferry service be stopped. A small boat service was being developed along the coast facing Sweden. Local training would be accelerated. Secret arms dumps would be set up all over the country.

Turnbull proposed that the next stage of operations should be co-ordinated by a field committee including Muus and Fog, working in collaboration with the Princes, shortly to be succeeded by the army officers of 'the Little General Staff'. He thought Fog would be particularly important because of his close association with the Communists, who had been responsible for much of the early sabotage, and that, to avoid overlapping, there must be division of labour between SOE and the army officers. He paid a tribute to the excellent work done by Flemming Juncker and his friends in reconnoitring key positions in Jutland, where they had enlisted the help of factory managers, local officials, and controllers of public utilities. Nevertheless, he feared it might all be wasted if, as seemed possible, the Little General Staff became so powerful that it could run the resistance single-handed.

He need not have worried. Any danger of serious competition from the army, if it ever existed, was effectively removed by the events of 29 August 1943.

It might be thought that the disbandment of the armed forces, the tame surrender of their arms, which could have been denied to the Germans, and the establishment of absolute Nazi control, would create insuperable problems for the Danish resistance movement. The position was saved by the formation of the Freedom Council (*Frihedsrad*), something of a miracle. Although the central administration continued to function, the removal of the government left a vacuum, and the varying political shades of resistance coalesced into a group which saw itself as doing the work of the elected government. The six founder members were Fog; Frode Jakobsen, who in 1941 had founded *Ringen* (The Ring), a nation-wide clandestine organization; Aage Schoch, former editor of *Nationaltidende*, who had been active on subversive propaganda; the Communist Børge Houmann; Jorgen Staffeldt, of the Danish Unity Party; and, representing the Free Danes, Erling Foss who had worked closely with SOE since 1941, and was later replaced by Professor Ole Chiewitz, who had been in charge of the Danish ambulance unit in Finland in 1939.

The fears of some politicians, shared by a few Danes who were aware of the activities of the Council, that the Communists might seek to dominate it in the hope of gaining power post-war were allayed by the support it received from Christmas Møller in London and the British authorities. It soon became evident that it was working for the good of the country as a whole and not for one political faction. Ultimately its membership increased to thirteen, including SOE's chief representative in Denmark. There was academic discussion as to whether Muus and his successor Ole Lippmann were technically members. The essential point is that, whatever their status, they attended Council meetings. They could put across SOE's point of view and convey to London their interpretation of Council policies.

It is a curious fact that it was not the Communists who posed a threat to the efficient running of the resistance machine, but the army officers. In Denmark, as elsewhere, the Communists had refused to have anything to do with the 'the imperialistic war' until 22 June 1941, when Germany invaded the Soviet Union and they became active overnight. BOPA (from *Borgerlige Partisaner*) initially comprised fifteen men who had fought with the International Brigade in Spain. They financed their activities by stealing, and thanks to bad security many were arrested. Houmann reorganized the survivors into two groups, neither of which knew about the other. Altogether 500 men served with BOPA during the war, thirty-one losing their lives. They carried out over 1,000 sabotage operations, and liquidated twenty Nazis and Danish informers. Ole Lippmann has given credit to the Communists for their contribution:

> In the field in Denmark we were working extremely well with our communist friends. We have different ideas as to the future, but we quickly agreed that we were shooting at the same enemy, and we went out of our ways to avoid episodes which might have created a split between, for instance, the Holger Danske Group and the BOPA Group. I personally had the closest touch with the leading communists Borge Houmann and Alfred Jensen, and have never been in doubt that they were loyal, nor that they wanted a free election after the war to fight for as many votes as possible, which after all is legitimate.

Holger Danske was a well-organized sabotage group which came into existence in 1943 and successfully carried out a great many sabotage operations, working very closely with SOE. (Holger Danske was a mythical Viking who slept in Kronborg Castle at Elsinore and awoke to save Denmark when she was in danger.)

While the Communists were thus playing an important part in the resistance, the army wanted to keep at a distance from the movement. It was, of course, hindered by loyalty to the elected government, although it was in cold storage, but it was also influenced by the hope that it would be possible to survive intact until the end of the war, so that when liberation came it would be in a position to dominate the post-war scene. Neither the Princes, in spite of their half-hearted assurances to Ebbe Munck in Stockholm, nor the later O Groups, were ever involved in serious sabotage. The army's bare-faced attempt to take over the 3,000 Husqvarna machine guns bought from Sweden, for which Ole Lippmann had begun to raise money in 1943, was not intended to equip it for a full-scale assault on the occupying forces—at least not until it was clear that the Germans were on the way out. The Freedom Council had no doubt that arms, from whatever source, should be equitably distributed among all members of the resistance proper. If there was to be discrimination, it was in justice the old-established resistance groups, including the Communists, which had early set their sights on victory within Denmark, that should be favoured, rather than the Little General Staff, with their eye on post-war benefits.

The resistance side of the Council's business was first handled by its M Committee of five: Stig Jensen, Ole Geisler, Jorgen Staffeldt, Svend Wagner, and Captain Aage Hoiland Christensen representing the army, who was anxious to integrate officers in the resistance but was removed by the Little General Staff. Flemming Juncker was an associate member. Of these men, Jensen, Geisler, and Juncker had been working with SOE from the beginning. In the beginning of 1944, when it was believed that an Allied invasion of the country might be imminent, the M Committee was renamed K and became more military in character, being intimately concerned with the direction of the resistance in the country's six regions. There were now representatives of the army, navy, police, and SOE, with a Council member as Chairman.

Muus was summoned to London in December 1943 to give a first-hand account of the state of affairs after the German take-over, and to receive briefing for the next phase of the campaign. SOE explained Allied plans for D-Day, so far as it knew them, and his future role as its principal representative in Denmark. The *Table* organization would be commanded by Ole Geisler, and the new 'resistance groups', the successor organization to the secret army, which retained the code-name *Chair*, by Preben Lok Linblad. Muus would co-ordinate the activities of the two bodies. He would ensure that the valuable flow of intelligence, now provided by Svend Truelsen, continued uninterrupted, and that propaganda material from Britain was efficiently disseminated. Above all it was essential that the resistance movement should be decentralized and take its orders region by region direct from London. He must not become involved in politics. Anything with political undertones should be referred to headquarters for consideration. 'If you fail to carry this out in letter and spirit, it will greatly undermine your authority in the field, and the trust we have placed in you.' According to his instructions he was in sole charge of SOE-sponsored resistance, and, since there was no mention of the Freedom Council or its M Committee, SOE saw him as sole commander of the movement, taking orders only from the Allied high command. That this was the SOE view was later confirmed by Hollingworth.

The Chief of Staff to the Supreme Allied Commander (COSSAC), Lieutenant-General Sir Frederick Morgan, had already decided to exercise operational control over SOE activities in all the countries with which he was operationally concerned, including Denmark and Norway. Although there was no possibility that the Allies would invade Scandinavia, the instructions to Muus had to envisage the possibility—otherwise they might provide a clue to Allied intentions which German intelligence would pick up. He was, therefore, given alternative instructions. If Denmark was invaded, *Table* and *Chair* would concentrate on impeding the movement of German troops and attacking fighter and light bomber aircraft on the ground. If she was not invaded, they would disrupt German lines of communication—important for bringing reserves from Norway—and tie down German troops within the country. If the occupying forces surrendered, patriots were on no account to harass them. Their job then would be to maintain order, and prevent pillage and marauding. Although it is now obvious that these even-handed instructions were inevitable, and were almost certainly part of the general deception plan to which SOE was not privy, Muus seems to have taken it for granted that Denmark was to be invaded. This is the impression he gave to his fellow-countrymen on his return from London in January 1944. According to Frode Jakobsen, Muus 'returned with a message that the expansion of the Danish resistance groups should be complete, ready for battle, from March 1944. We understood that from then on an invasion of Denmark was not improbable.'[2] Muus was unwittingly playing the game of the London Controlling Section, the COS deceptionist organization.

During Muus's absence in London there was a marked slowing up of activity in the field; and his own contribution was almost terminated when the aircraft taking him back to Denmark was shot down, but without casualties, on 11 January 1943. He divided the crew into two groups, one of which he led to safety, and sent to Sweden. The others were taken prisoner. Muus now showed a remarkable and uncharacteristic disregard for security, which is difficult to explain. He continued to use his code-name *Jam*, although it was compromised, and was severely reprimanded by London. He incurred the censure of Turnbull by sending photographic messages to Stockholm in the original 35 mm tins marked 'undeveloped film', with people's names written on them, which made nonsense of the security arrangements. 'Even in our wildest days we have never used such a system. Surely if we are to use films for messages, they should look as little like films as possible.' London agreed, and remonstrated with Muus, who continued to use the original tins, and even sent them by post so that they fell into the hands of the Swedish censorship, and had to be retrieved through the good offices of Nordentoft. Turnbull became so exasperated with this total disregard of security that he reprimanded Muus direct: 'In all my experience of this work I have not known such a series of blunders, and I must ask you as a friend to remedy the situation . . . Please do your best to stop the rot.' Turnbull was in turn reprimanded by London for exceeding his authority, and very reasonably replied that headquarters' failure to make Muus toe the line fully justified his intervention in a situation where SOE's secrets were being handed on a plate to the Swedish authorities, and who knew to whom thereafter.

The main problem facing Muus was how to recruit and train the new resistance groups in good time for their presumed imminent action. A suggestion by the M Committee that the reservists on whom the Princes had supposedly been relying should become a secret army controlled by them was rejected on the ground that it would mean centralization which might compromise both *Table* and *Chair*. Far better for the men to be used to build self-contained resistance groups. It was contemplated that regional organizers, instructors, and W/T operators would be trained in Britain, to be dispatched to Denmark round about D-Day. Men would be found with specialized knowledge of Danish coastal waters—fishermen, seamen, pilots, tug-masters—who could be enlisted at short notice in the event of an Allied decision to invade. (Again, this savours of part of the deception plan to hint to the Germans that Denmark would be invaded.) Ideally, a national organizer would be found in Britain, where he could be made familiar with SOE's specialized training and methods, although he would be less familiar with local conditions than someone recruited within the country. That there was still an element of unrealism in SOE's planning is witnessed by the proposal that large numbers of agents should be trained in Britain to serve in the new operational groups. There was no time to train them; and, as was belatedly recognized, at the time of D-Day only a few could be sent in by air, probably no more than twenty.

A consequence of the decision to build up resistance groups was that the supply of sabotage material from Britain gave way to the delivery of arms. The *Table* organization was told to go slow on sabotage, partly to conserve stocks of explosives, and partly because it was believed that, if the saboteurs kept a low profile, the Germans would become less vigilant and the reception of arms would be easier. Muus agreed to this change of programme while he was still in London; but on his return to Denmark he pleaded for some sabotage simply to keep *Table* active. At the end of April he was authorized to carry out three or four sabotage acts weekly. The next month he was told, in accordance with an order from the Allied high command, to continue on a rising scale, and to send London reports of all major activities.

The planning directive for SOE activities in Denmark which was evolved in March 1944 showed how completely the idea of a secret army had disappeared. It accepted the *Table* organization as a going concern: 'closely knit and with about 1,000 trained men fully conversant with the technique of reception and distribution of supplies, and with demolition and incendiarism.' It would function as hitherto, controlled from London through a chief organizer. The duties assigned to the new *Chair* were virtually the same as *Table*'s. Twenty thousand men were available, the best of whom would be recruited in small independent resistance groups by a member of the *Table* organization. These groups would work under liaison officers sent from London, and would be responsible for most of the activity on D-Day and after. *Table* would receive and distribute their supplies and would provide their W/T operators.

The tasks of the two bodies were defined. *Table* would concentrate on major activities. *Chair* would engage in minor sabotage, more widespread and therefore equally important. *Table*'s targets would be ports and shipping, power stations, and radio stations, which called for highly trained specialists. They would not need large quantities of explosives, but their delivery must have priority in view of the importance of the targets. *Chair* would be in two sections. *Chair R* had 400 teams of six men, the best qualified in the region, representing all political parties. Their objectives would be railway lines and signal boxes (rather than heavily guarded railway stations), telephone lines (rather than exchanges), and power cables (rather than power stations). *Chair F*, absorbing all other volunteers, would take care of minor targets—blocking roads, attacking vehicles, petrol and oil supplies, removing signposts, raising false fire alarms, and generally hampering troop movements. It was accepted that the nature of the terrain afforded little scope for permanent damage, but tyrebursters, broken glass, and felled trees could all be used for temporary disruption.

Table's instructions would be issued through film messages—well tried and successful when properly used—and W/T; but something new had to be invented for the *Chair* parties, too numerous to equip with W/T or channels for photographic instructions. Each region was allotted its own code words to be broadcast by the BBC. One word indicated the type of target, a second the nature of the attack. If Region V was required to attack German communications, *Chauffeur* (meaning communications) would be broadcast, followed by *Driver* (meaning 'stand by'), or *Artist* ('do utmost damage'), or *Haberdasher* ('do damage repairable within a week'). If the target was to be protected from a scorched earth policy, the signal would be *Chauffeur–Motorist*. This highly flexible system meant that orders could be changed at short notice and that the nature of the campaign could be varied from region to region. This applied only to *Chair R*. *Chair F* would simply be told to go into action, the signal being repeated every three days. If the signal was not repeated, the action had been called off. The code signals would be issued through the M Committee, which would then go to Sweden 'to eliminate all centralization in the field'. Thereafter Allied high command orders would be issued through the BBC direct to the regions.

All this was put to the representatives of the Freedom Council at meetings in Stockholm in April 1944. Hollingworth came from London; and OSS was represented by George Brewer, head of its 'Westfield' Mission in Stockholm. Frode Jakobsen, Arne Sorensen, and Erling Foss represented the Freedom Council. Ebbe Munck and Nordentoft also attended some of the sessions. Muus should have been present but was laid low in Denmark with a bout of malaria, a legacy of his career in Africa. His place was taken by Ole Geisler. The main problem was that, while SOE, with the backing of SHAEF, wanted the six regions in Denmark to be self-contained and take their orders from London, the Freedom Council wanted to keep control of the whole country in its own hands. SOE saw this as dangerous centralization which would lead to disasters of the kind suffered by *Milorg* in the early days in Norway (see Chapter 8). In spite of its misgivings, the Freedom

Council representatives seemed to accept the Allied proposals, and returned to Copenhagen taking the directive with them.

Hollingworth told Muus that the British team was delighted by the co-operation shown by the Danes. In particular SOE's own man, Ole Geisler, had shown great diplomacy. The success of the conference was largely due to him. It now remained for the Freedom Council to ratify the agreement entered into by its delegates. Ever the diplomat, Hollingworth added that everyone had spoken very highly of Muus; and further that he had been made more aware of the difficulties he had had to overcome in the field.

Surprisingly, Muus responded with a catalogue of personal grievances. Major Ray of the Stockholm Mission had undermined him in conversation with Geisler. Frode Jakobsen, who brought the directive back from Stockholm, appeared to think it was for his eyes only, and further that Muus's role as *Table* leader was finished. Worse, Jakobsen was financing the collection of intelligence without reference to Muus, which neither he, nor Truelsen, in charge of intelligence, could understand. People had got the impression that London had lost confidence in him. He therefore tendered his resignation, and recommended that he should be succeeded by Geisler.

Hollingworth, accustomed to dealing with prima donnas in the field, instantly replied in a long and carefully drafted telegram. The directive had been entrusted to Jakobsen as representative of the Freedom Council, whose presence Muus had fully approved. It was never intended that he alone should see it—he was no more than a courier who brought it to Denmark—and, if Muus wanted a copy, one would be sent. Every decision at the Stockholm meetings had the full approval of Geisler, who by now must surely have put Muus in the picture. In any case, the directive did not tell the whole story—the most important future moves would be sent to him direct by W/T as each stage of the Allied masterplan was reached. Headquarters were 100 per cent behind him, and had no doubt that his diplomacy would overcome all difficulties. 'Softly softly catchee monkey. Regards. Hollingworth.'

Muus replied: 'Thanks for the vote of confidence.' His gratitude might have been tempered had he seen Hollingworth's file note recording the care that had gone into drafting his soothing telegram: 'The exact ratio of fact, soft soap, commonsense, humour required was very carefully weighed up.' He believed the threatened resignation was

the natural reaction of an ambitious young man who firstly is annoyed at having missed what turned out to be the most important negotiations and secondly cannot reconcile himself to the necessity of dropping many interesting activities that he previously personally took a hand in.

(He might well have added that Muus probably deprecated the fulsome praise meted out to his deputy for his performance at the negotiations.) Muus was given an assurance that, while the importance of the Freedom Council was now fully accepted by the Allied high command, and while there was no wish to dissociate it from the resistance movement, he remained the leader of SOE's sabotage organization, and London's representative on all matters. In passing this on to Stockholm, Hollingworth stressed that the whole episode was no more than a storm in a teacup.

When the Freedom Council examined SOE's plan in Copenhagen, it was at first bitterly disappointed by what it saw as a transfer of its executive power to the Allied high command; but eventually it agreed to accept it. This may have been done tongue in cheek, in the knowledge that, if it wanted to ignore some aspect of the plan, SOE could hardly force it to come into line. Thus, while it was a key part of SOE's proposals that the M Committee should withdraw to Sweden after the arrangements were fully in train (any who remained in Denmark might be captured and compelled to blow the whole plan), the Freedom Council had no intention of complying with this provision.

The independent line followed by the Council (natural in the circumstances) meant that in effect it was running in parallel with SOE, joint managers of the resistance movement, with less opportunity for the close consultation that joint management calls for. SOE had provided the W/T network, vital for the success of the movement, and a good many operators; was responsible for the supply of arms and equipment, and largely for the arrangements to receive them; it maintained liaison officers in all the regions; and sent in a few teams of specialist saboteurs. The Council's M Committee, with SOE's chief representative as an honorary member, effectively controlled activities in the regions. How this joint management could have coped had there been a 'French situation' in Denmark—had the resistance been called on to give all-out support to an Allied invasion—it is hard to say. It was not put to the test.

The great increase in sabotage at the end of 1943 led the Germans to launch an anti-sabotage campaign with the theme that it was the duty of every good Dane to preserve a correct attitude towards the occupying power; and that, although sabotage had no more than a nuisance value, it was sowing the seeds of disaster for Denmark's economy post-war. The campaign was aided by the activities of the so-called Schalburg Corps (Danes originally recruited to fight on the Eastern Front and now an auxiliary force in Denmark), which was required to carry out pointless acts of sabotage in an attempt to discredit the genuine resistance movement. It was feared for a time that the campaign was having an effect and that sabotage might come to a halt through lack of popular support, but in fact as D-Day approached the resistance movement regained its momentum.

The first six months of 1944 saw intensive activity by the Germans against the resistance, and there were many arrests, the 'snowball' effect of which was sometimes disastrous. In March Muus reported to London that two agents had given away many colleagues, almost gratuitously:

> Believe it or not, the only form of torture to which they have been exposed is the lack of cigarettes. A number of times the Germans have given them a cigarette, lighted it for them and taken it out of their mouths at once promising them the rest of it if they would squeal some more names. This they have done readily . . . The Germans despise them for their behaviour.

Muus claimed that one of these men already had '40 good Danes on his conscience'.

It is quite impossible to do justice to the bravery of all the saboteurs during this period; but one attack against Odense railway station in March 1944 typifies the skill and daring that went into the campaign. Forty-three men in seven groups carried out simultaneous attacks on different targets to delay through rail traffic from Zealand to Jutland, although the station was guarded by thirty Germans, six or eight of whom patrolled in pairs night and day, and although the danger was compounded by the large number of saboteurs operating at the same time. Four more guards were stationed by a tunnel carrying the main highway under the railway. A station employee responsible for detecting sabotage helped by providing a plan of the station, and by carrying out a preliminary reconnaissance.

With the aid of moonlight and the dim lighting of the station, four men attacked a gantry carrying signals. As one end was close to the guard house, they had to work on the other end, where they pressed 6 lb. of plastic explosive into the girders. Another four dealt with a second signal gantry adjoining a signal box, after holding up the signalman.

Three more went in through the main entrance with 5 lb. of plastic rolled into sausages and wrapped round their waists, to put three water towers out of action. Two planted their explosives successfully, but the third was eyed suspiciously by a guard and withdrew without placing his charge.

Six were told off to attack the principal water tower. They had hoped to plant their explosives inside, but could not open the door. They had to lay their charges against it, where they would be less effective. An attack on the turntables also had to be modified. It was impossible to locate the manholes giving access to the space underneath the turntables, where an explosion would have been devastating, so the charges were jammed on the surface. Another team of six was equipped to demolish up to seven locomotives in the round house, but to their disappointment found only two shunting engines, on each of which they placed 2 lb. of plastic. Yet another group dealt with an engine standing in the open. The seventh group alone failed to accomplish anything.

All the charges were detonated by time pencils pressed simultaneously at 8.30 p.m. with a 30-minute delay, which gave the attackers, who planned to be out of the station by 8.45 p.m., fifteen minutes' grace. When the nine separate explosions went off within a few seconds of each other, it seemed as if the whole station was being blown to bits. Most of the attacks were completely successful. One signal gantry was out of action for a month. The second had not been repaired at the end of the war. The cylinders of the two shunting engines were blown off. The signal box was completely demolished. The turntable was only slightly damaged because of the failure to plant the explosives underneath, and was operating again within a few days. The most serious result of the attacks was the destruction of the facilities for watering engines. Simultaneous attacks on water towers elsewhere on the system meant that, over the whole Fyn main line, engines had to be watered from town supplies, a slow and inefficient process.

On 30 April 1944, with D-Day just over a month away, the saboteurs in Copenhagen took time off their normal activities to honour the birth of a daughter to the Crown Princess. A group from the Holger Danske organization simulated a 21-gun salute (planned many months earlier when it was announced that the Crown Princess was expecting a child) by exploding twenty-one bombs equipped with appropriately varied time delays—a splendid gesture showing that the morale of the resistance was still high.

8

The secret army to D-Day: Norway

THE potential importance of the resistance movements in the occupied countries of Europe depended on the route ultimately to be chosen by the Allied high command for its return to the Continent; and it was obvious early in the war that neither Norway nor Denmark would be the point of re-entry, although many Norwegians believed their country would be chosen for invasion long before the Allies landed in France in June 1944. It followed that their resistance movements in general, and their secret armies in particular, would not be required to play a major role in the defeat of the German army, and their own liberation. This did not mean that SOE could afford to ignore clandestine activities in these countries—to give no support at all might provide the enemy with a clue about the eventual point of return to Europe; and, without arms and equipment supplied by SOE, it would be difficult for them to tie down German troops and keep reinforcements in due course from the battle fronts in France. As it turned out, the secret army in Norway, *Milorg* (from *Militaerorganisasjonen*—military organizations), and *Danforce*, in effect the lineal descendant of Denmark's secret army (*Chair*), were not called into full-scale action in the field although they did carry out policing and anti-scorch operations at the end of the war. With the benefit of hindsight it is clear that both countries would have made a more effective contribution to the Allied effort had all the resources of their resistance movements been thrown into sabotage, at the expense of their secret armies. This would certainly have led to even fiercer retaliation and reprisal; but wars are not won without casualties. The secret armies in Norway and Denmark were primarily intended to control the occupying forces if they refused to surrender when the war was won in Germany. Unlike the sabotage groups, they were not part of the war of attrition in enemy-occupied countries; but at least in Norway the secret army provided a means of keeping patriotism alive during the years of occupation.

The first resistance groups in Norway were formed, immediately after the occupation, by young men filled with a sense of frustration anxious to prepare for the day when they would drive out the enemy—

for example, members of sports clubs who met to practise with the few weapons they could find. Tore Gjelsvik's experience is typical:

> In the autumn of 1940 I joined 8 or 10 good friends in starting the first *Milorg*-squad in my home neighbourhood. We were active sportsmen, with experience from the Scout movement and life in the woods. Several of us had taken part in the fighting in the spring, and there was no lack of courage and enterprise.[1]

In the words of Jens Christian Hauge, leader of *Milorg* for most of its existence: 'The men joining *Milorg* did so because they declined to accept that the war was over with the capitulation in 1940. Their ambition was to be good soldiers, irregular yes, but definitely part of the Norwegian forces.'[2] The scattered groups coalesced into larger formations, and in Oslo two officers (Major Olaf Helset and Captain John Rognes) who had taken part in the brief campaign against the German invaders arranged for others to give more professional instruction. These formations gradually developed into *Milorg*, recognized by the Norwegian government in Britain towards the end of 1941, and brought formally under the Norwegian high command.

The absence of efficient communications made it very difficult for SOE to get a clear picture of what was happening in Norway. In October 1940 Malcolm Munthe in Stockholm provided a list of men in Norway 'working secretly to keep alive the spirit of resistance', including 'a charming dapper little sportsman of some 60 autumns who begged me to send him out to kill Hitler or Quisling or give him a plane in which to do a death dive over some important objective'. In spite of Munthe's assurance that this man was not a lunatic, SOE cannot have been much impressed by this picture of potential Norwegian resistance. More convincing was his statement that 'There are in Stockholm a large number of loyal swashbuckling enthusiasts who are entirely unusable.' In January 1941 one of Munthe's agents brought back from Norway information about a secret organization there, but refused on security grounds to give any more than a very sketchy outline.

Another agent brought a report of a group of twenty-five men with five machine guns and eighteen rifles, who had hidden skis, sledges, axes, compasses, binoculars, maps, and tools in mountain huts; but it was admitted that few were as well equipped and as well prepared. An examiner in the Royal Patriotic Schools at Wandsworth, where refugees were interrogated, was so struck by the numbers who spoke of military organizations all over Norway that he took it upon himself to propose that SOE should co-ordinate their activities. He thought it impracticable to set up a headquarters in Britain and suggested Sweden instead. Tennant in Stockholm had much the same idea. He knew of several officers in Norway trying to work out a scheme for concerted action 'for use if and when we land'. He thought SOE should give a lead which a central organization in Norway could not safely do, and, if London agreed, he would ask group leaders for suggestions. He was told that SOE would sponsor any responsible organizations big enough to be useful—but they would have to be controlled from Britain.

These reports reflected accurately the true state of affairs—an organization very much in embryo. Yet a memorandum sent to King Haakon in June 1941 via the Shetland Bus Service (see Chapter 5) describes an organization seemingly in full working order, with a supreme military council of two members, and five active groups in Oslo, Lillehammer, Kristiansand, Bergen, and Trondheim. The degree of organization varied from place to place. Training could be carried out only in the evenings when men had finished work. Arms were not needed at this stage since there was no intention of engaging in sabotage. 'The idea behind our organization is to make available a military machine to support an internal government sanctioned by the King . . . which will decide if and when the military machine will function.' So far so good. It is when the memorandum describes the military machine in detail that it may seem to wander into Wonderland. It has 14 branches: organization; pioneer; air; Red Cross; supplies; transport; maps; propaganda; internal messenger service; finance; intelligence; courier; communications Norway–Britain; and staff matters—a dream of clandestine enterprise which must take a long time to become a reality.

That SOE accepted all this uncritically is witnessed by a paper of 10 June 1941 which complacently states that there exist in Norway a number of indigenous anti-Nazi organizations of a semi-military character formed district by district, which by the end of the year will have a membership of 20,000 covering the whole country. Ammunition dumps will be established in each district. It is then surprisingly claimed that 'The co-ordination and also the training and equipment of the various bodies is being directed by SOE and large preparations are in hand which it is intended to carry forward immediately there is sufficient darkness to resume regular communication by boat . . .' SOE prepared a reply to the memorandum to the King in which it sought to dispel the idea that anyone in Norway would decide when to go into action. 'Eventually a day will come when, with the opening of a British offensive on the continent of Europe, the war against the German invader will break out simultaneously in every land which is invaded.' (The SOE draft had said 'every land which he has conquered' which was amended by a sensitive pen on the Norwegian side.) 'It must be clearly understood that it will not be for any governing body in Norway to decide when the machine operates. This can only be done by Norwegian officers under the direction of the British General Staff.' It is not clear whether this directive was ever sent to *Milorg*, but at least it reveals SOE's attitude to the secret army at this stage—as eager to believe in its great potential as it was to believe in that of the Danes', *and* to regard it as part of its own organization. SOE did, however, stress that the army should go into action only in support of an Allied landing, otherwise the Germans would destroy it and take punitive measures against the people at large.

In July Dalton sent Churchill a plan for special operations in the next twelve months in which he said he would have in Norway a secret army, 19,000 strong—a figure plucked from the air. The JPS, to whom the plan was referred, confirmed that, if the British army was to return to the Continent, the help of local patriot forces would be essential.

> Our expeditionary force would be enormously assisted if part of these patriot local forces were previously secretly organized and armed so that they could be employed effectively in interrupting communications, attacking aerodromes, and other diversionary operations, from the commencement of our attacks on the continent.

But secret armies must be encouraged only in areas where an Allied offensive was practicable, which meant Holland, Belgium, and northern France. In Norway there would at best be no more than subsidiary operations, but some arms should be sent, if only to keep up morale. The COS agreed, but stressed that the needs of the regular forces and secret armies elsewhere must come first.

Meanwhile the embryo organization was suffering setbacks. The Gestapo had discovered what was going on, and early in 1941 had arrested one of the two leaders. The other, Captain John Rognes, escaped to Sweden and eventually to Britain. Many of *Milorg*'s more active members were arrested, but replacements came forward and losses were made good. Unfortunately, in the course of 1941 the groups looked more and more to the central leadership in Oslo, *Sentral Ledelse* (SL), for direction, playing into the hands of the Gestapo, who in the autumn made many arrests in Oslo, Bergen, and Trondheim. The dangers of centralization were compounded by the attitude of some members. 'They were too naïve. Because a friend had been a friend he could never be a traitor and informer.' At this time an insular minority view in SOE was that things would be better 'if we, the British, would only take bodies of different nationalities and work them ourselves and keep them away from their so-called national governments'; but the co-operation of the Allied governments was, of course, essential.

In November 1941 SOE began to question its relationship with the growing military organization, prompted by the arrival in Britain of some leading members. *Milorg* wanted day-to-day co-operation with the Norwegian high command; while the Allies wanted it to keep in step with their strategic policy for Europe as a whole. It was, therefore, necessary to define the nature of subversive operations for Norway. It would be left to *Milorg* to intensify passive resistance. The secret army must have in each district a well-trained force for the reconquest of the country; and SOE must send in arms and equipment right away. Sabotage, left to agents sent in from Britain, must be aimed at important strategic objectives. Only in exceptional circumstances would it be carried out by the secret army. It was accepted that the Anglo-Norwegian Collaboration Committee (ANCC, mooted to maintain a link on special operations between the two governments, but not yet set up) would have to be kept informed about SOE's activities, but there might be occasions when the Norwegians must be kept in the dark, for example when SOE personnel were loaned to Combined Operations for an attack which had to be kept particularly secret. SOE continued to be worried about centralization. One solution was to send leaders from Britain entirely dissociated from the organization in Oslo.

Early in 1942 there was still uncertainty about the respective roles of SOE and *Milorg*. Their objectives, or at least the means of attaining them, did not coincide. *Milorg* saw itself as a secret army which might be jeopardized by large-scale sabotage. A directive of January 1942 'did not envisage guerrilla and sabotage actions as long as the Wehrmacht remained unattacked in Norway'.[3] SOE accepted the need for a secret army whose claims ranked well below those of the corresponding organizations in France, Belgium, and Holland; but saw current sabotage as part of the campaign against Hitler. There were other differences. An officer who left Norway in January 1942 complained that SOE was a competitor and that *Milorg* should be subordinate only to the Norwegian high command. He was assured that there was no intention to run a parallel organization. SOE continued to worry about centralization and considered it important to strengthen W/T communication with the *Milorg* districts to avoid the need for dangerously close ties with Oslo. It was hoped that the ANCC would improve co-operation, but there was still a long way to go. Ignorance in London of the nature and potential of the secret army continued to be worrying. The cell system on which it was based meant that the rank and file who came to Britain could not give a comprehensive picture; and senior leaders rarely left the country.

In June 1942 Wilson, head of SOE's Norwegian Section, examined objectives in Norway. The main problem was 'the absence of information concerning high policy'. You could not plan for The Day when you had no idea when The Day would come. If the secret army was ready too soon, morale would disintegrate. On the other hand, the resistance forces might be called into action before they were fully trained. 'It is true that secret organizations must work in the dark, but some starlight would be most helpful . . .' Here the Norwegian Section was in a cleft stick. Planning would be simplified if it knew when the secret army would be needed; but to reveal the date (or that it would never be called on) would be suicidal. The information would be in Berlin in a matter of hours. But because of the uncertainty many were impatient, indulging in rash overt actions, so that supposedly secret groups became known and blown. A special problem in Norway was the simplicity of the people, who had a childlike confidence that no one would give them away. They carried out their secret work in public, using the telephone much too freely. Further, there was the arrogance of the military class who assumed that they alone could defeat the enemy, although they were as much amateurs in the modern science of total war as the amateurs they criticized. In Wilson's eyes another adverse factor was the low opinion held by many within Norway of the government in exile, as well as continual complaints about the performance of the Norwegian Legation in Stockholm, which made it difficult for the Norwegian authorities in Britain to unite their loyal compatriots both at home and in exile. Morale was at a very low ebb. The Legation was likened to the headquarters of a routed army.

SOE's role must be to steer a safe passage among all these shoals, so as to create the maximum resistance to the enemy in the event of an Allied landing. A *sine qua non* was to keep groups established in Norway by SOE totally separate from the secret army—a policy which the ANCC had just endorsed; but what was sound in principle—don't carry all your eggs in one basket—was not easy in practice. Norway was a large country with a small population, and the arrival of anyone from outside was bound to arouse unhealthy interest. SOE agents had come in contact with members of the secret army, sometimes wittingly, sometimes not, with the danger of compromising both sides. Above all it was essential to ensure that the secret army would act only on orders from the Norwegian commander-in-chief in Britain, and not from Oslo headquarters, where Allied strategical requirements could not be known. Ideally the number of W/T channels would be doubled so that London could communicate separately with the two organizations.

Examples of line crossing which Wilson deprecated included the case of *Anvil* (F. B. Johnsen) sent to Lillehammer in Opland, with explicit instructions to keep well away from the local *Milorg*, and to build his own guerrilla group. When *Milorg* headquarters in Oslo heard about this, they asked London to make *Anvil* subordinate to them. SOE saw this as an attempt to return to the discredited and potentially disastrous practice of 'leading all strings into a centre in Oslo'. Again, SOE's agent *Crow* (E. K. Jakobsen) had been induced by the local *Milorg* group to put his W/T set at its disposal; and, when Jakobsen returned to Britain, *Milorg* simply took over the whole *Crow* organization. Thirdly, SOE's *Mallard* (E. K. Martinsen), in the Bergen area, suffered the same fate. It was becoming impossible to avoid the hopeless entanglement of the two organizations. There must be a definite agreement that *Milorg* was a latent body to go into action only when there was an Allied landing in Norway. If it was deemed necessary to operate guerrilla bands before then, they must be SOE's responsibility. If *Milorg* wanted earlier action, it must detach small groups to operate quite separately, which would be equipped and trained by SOE.

In coming to these conclusions Wilson had the benefit of a report by Captain Schive, a senior *Milorg* officer who had come to Britain, and was sent back to Norway by General Hansteen, the Norwegian commander-in-chief, to determine just how far the secret army had progressed. Schive's findings were that the leaders had become much more security-minded and that they had 'engineered a defence in depth which the enemy will find it virtually impossible to penetrate'. A central council of four—a soldier, a doctor, an engineer, and a trade unionist—were directing *Milorg* in south Norway, now divided into fourteen districts, in two groups of seven, each in charge of an 'inspector'. The total strength of the organization was put at 25,000 of whom 12,000–15,000 were armed, the rest being known as 'pioneers'. One reason why the earlier leaders had been vulnerable was that they had been known to be staunch opponents of occupation; but the current practice was to select men who appeared to be collaborating. The police and security services were well represented and gave advance warning of trouble. Wilson was impressed by this account—perhaps in spite of himself—but he could not yet give *Milorg* his unqualified approval. The report 'paints a picture which is brighter than was previously thought. The colours have not been laid on thick, but at the same time the picture is not ready for exhibition. It is at present about three-quarters finished.'

Gubbins was rather more impressed. He told CD that SOE had been trying for the last six months to find out exactly what *Milorg* amounted to—'how much was real, and how much just talk'. There had been early setbacks 'owing to various stupidities and bad organization, but it has learned from its mistakes and seems to be a really good conspiratorial organization'. Gubbins directed the Norwegian Section to work as closely as possible with the secret military organization. More W/T links must be provided, instructors and arms sent in. The Section must press on with plans for the sabotage of communications, but there must be no railway sabotage without specific instructions from headquarters. *Insaisissible* sabotage must be greatly increased, and—an echo of the problems with Denmark (see Chapter 6)—*Milorg*, despite its objections to sabotage, should be required to attack one worthwhile target, to test its ability and willingness to co-operate.

Wilson, who three months earlier, with the support of the ANCC, had argued that SOE must go it alone in Norway, now dutifully followed Gubbins's line. He wrote: 'After mature consideration, and in the light of events, I am confirmed in the opinion that has been gradually forced on me since I was placed in charge of the Norwegian Section, that SOE long-term policy in Norway demands drastic revision.' (A curious opening gambit—Gubbins's month-old directive laid down that SOE policy in Norway should be drastically revised.) Wilson was influenced by the poor performance of missions sent into Norway. Whatever the reasons—and bad luck may have been the least of them—they had failed to establish a foothold. *Arquebus* (Bernhard Haardvardsholm) had been lost at sea when returning to Britain to report. Erik Hansen of *Archer* had been taken prisoner. In Hordaland Martinsen and his second in command had also been taken prisoner. Jacobsen survived only four months before falling into the hands of the enemy in Akershus. There was a major disaster when *Penguin* (A. M. Vaerum) and *Anchor* (Emil Hvaal) were cornered by the Gestapo within a few days of landing at

Nesvik. The pair were hiding in the loft of their contact's house at Televaag and defended themselves, killing two senior members of the Gestapo. Vaerum was killed outright. Hvaal, seriously wounded, was taken prisoner. He recovered and, after prolonged interrogation and torture, was executed in October 1943. The failure of the missions was bad enough, but the ferocity of German reprisals made many question whether SOE's intervention was doing more harm than good. Seventy houses, barns, and boatsheds in Televaag were razed to the ground. Eighteen Norwegians who had previously been arrested for trying to escape to Britain were shot.

This episode made the overall picture even more depressing. Of forty-seven SOE-trained men—organizers, sabotage instructors, and W/T operators sent in from Britain—at most only twenty-eight survived. Nineteen had either been killed, or captured, or were back in Britain. The balance sheet for arms and explosives was no better. Of 80 tons delivered in the previous twelve months, just under half had been captured.

These failures, however, were only part of the case for revising SOE policy in Norway. More important was that it was now obvious that any attempt to set up separate SOE and *Milorg* groups in the same area was detrimental to both, and doomed to failure. It was impossible for outsiders—that is SOE-trained Norwegians sent in from Britain—to avoid coming to the notice of the Gestapo, unless very close cover had been prepared. It was equally impossible for SOE to recruit better men than could the existing national movement 'embracing as it does all loyal Norwegians of any standing and worth in South Norway', the very existence of which would be threatened if it were made subject to foreign influence. The promise of loyalty and obedience to the King, through the commander-in-chief, was a token of its unity and strength.

A decision 'to stop forthwith any further attempts to set up any SOE organization, as distinct from, or even parallel to' Norway's secret army was timely. In future SOE would do its best to meet the requests of the Norwegian high command for the provision of agents, who would be sent to Norway only on the direct orders of that command, their operational instructions being worked out jointly by SOE and the Norwegian authorities. SOE would provide transport—by air or sea—for the chosen agents. 'It is to be clearly understood that they are to act in the capacity of the commander-in-chief's messengers, and are not expected to take command in any area unless specifically instructed to do so.' The only exception to these rules would be in north Norway where *Milorg* was not represented, and where it would be open to SOE to operate if it seemed expedient.

Malcolm Munthe felt that Wilson had gone too far in thus surrendering all initiative and planning to the Norwegians. SOE would now be no more than a tool in their hands. The Norwegian Section would have to rely on them for intelligence about affairs in Norway—which was inadvisable in the light of experience. Munthe recalled earlier misleading reports of 15,000–25,000 *Milorg* men in every province. In the event, however, the division of responsibility between SOE, the Norwegian high command, and the leadership in Oslo never became as clear-cut as these decisions suggested. There was some overlapping and friction almost to the end of the war—for example, the competition between SOE and *Milorg* in the *Antrum* area which was settled only at the beginning of 1945 (see Chapter 11).

The Norwegian high command was quick to ensure that the new arrangements were made known in Oslo. The leaders were told that from now on all activities would be directed by General Hansteen through them. The British contribution would be limited to supporting the organization in Norway. Any groups sent from Britain would subordinate themselves to the home leadership, provided it was represented in the area in question. This left SOE some freedom of movement, since *Milorg* did not claim to be seriously operating in north Norway. A directive from Hansteen to district leaders was still couched in terms of supporting an invasion, or 'taking over and securing power for the constitutional authorities' in the event of a German withdrawal; and it approved a formula for manning the fourteen districts. Each would have from two to four sections divided into two to four sub-districts, and so on through companies, troops, and finally, at the bottom of the pyramid, two to four parties. If the district leader opted for the minimum force provided by this very flexible formula, he would find himself in command of a mere sixty-four men. At the other end of the scale, it would produce a force of 4,096. In practice no district leader or subordinate commander could tie himself to such a rigid formula—the size of his groups and component parts must depend on local conditions, which varied from district to district. This suggests 'over-planning', an idea which is supported by some of the command structures. In District 14, for example, the establishment was impressive—eight departments with a chief for each. But there were in post only the Chief of Communications, the Chief of Pioneers, and the Chief of War Police. The posts of Chief of Transportation, Chief of Arms, Chief of the Medical Service, Chief of Supply Services, and Chief of the Civil Service were all vacant. The Norwegian high command took a more realistic view. It instructed *Milorg* to work on the simplest possible lines. Preparations must not be too detailed, and elastic enough to be changed at short notice.

SOE's understanding of *Milorg*'s problems now began to improve, thanks to visits to London by senior members of the organization. These men had to find their way across the border to Sweden, risking capture by the Germans and trouble with the Swedes. They then had to take their place in the queue waiting for the few seats on the air service from Stockholm to Britain. Alternatively, they could book a passage on the uncomfortable and potentially dangerous Shetland Bus. 'Certain emissaries' came to England in October 1942—convenient timing, as the Norwegian Section had just decided to change its policy—to report on the state of the secret army, and to plead for closer co-operation with London. There were further meetings at the beginning of 1943 when the *Milorg* representatives explained that increased enemy vigilance was making it more difficult for Oslo to exercise central control without endangering security, and asked that there should be more direct communication between London and the districts in Norway. Captain J. K. Schive invited Wilson and Munthe to a working dinner with the recent arrivals: Captain Osterras, who had been in command of the pioneers; Dr Ole Malm of the doctors' association; Arne Okkenhaug of the teachers' association; and Arthur Hansen, formerly secretary of the Secret National Organization, which represented all the civilian organizations opposed to the Nazi regime.

The Norwegians said there was a strong feeling in Norway that London had so far failed to understand the true position of the secret army; and they were under instructions to try to put across *Milorg*'s point of view. They now realized for the first time that Oslo equally had failed fully to understand SOE's position, and the many difficulties it faced, for example the impossibility of laying on unlimited air transport. The main need was still to increase the number of W/T stations in the districts to enable them to communicate direct with London, and thus minimize the volume of internal communication with its attendant dangers. The visitors also had discussions with Colonel Bjarne Oen of FO IV when it was confirmed that future operations from Britain should be organized direct with the groups in Norway, rather than through Oslo headquarters. But of course Oslo would have to be kept in the picture. A serious problem was the poor liaison between SOE's representatives in Stockholm and the Norwegian Legation there. There should be some form of joint organization in Stockholm.

After these meetings SOE and FO IV (who agreed that SOE had fully met the wishes of the Norwegian high command in the matter of co-operation) asked themselves how things could be improved. It seemed that *Milorg* was developing in a different direction from that originally intended. 'It has by degrees taken the nature of a positive fighting organization.' If it was to have any strength at all, it must be closely knit. The commanders must know their troops, and, more important, the troops must know their commanders. But the whole essence of a secret army was that the constituent parts must be kept separate and joined only by 'invisible links'. The earlier agreement that operations planned by SOE should be carried out through *Milorg* was not working well in practice. For one thing, the *Milorg* machinery had not worked fast enough, because of communications difficulties; and there was still the fear that, if SOE's enterprises were centralized through Oslo, they would suffer the fate that had befallen *Milorg* in the early days.

It was no easier for SL in Oslo to assess the efficiency of its district organizations than it was for SOE and the Norwegian high command to assess the efficiency of SL. The links between the leadership and the men in the cells at the bottom of the six-tier pyramid, if not quite 'invisible', could not convey much more than a very nebulous indication of the true state of affairs. If a 'party leader' claimed that the two, three, or four men he commanded were first rate and raring to go, there was no guarantee that when it came to the crunch they would be raring to go, or that they would be first rate. In the early days the leaders at all levels were doing little more than playing a 'numbers game', which could only become realistic as time went on and more men received training and arms. It was difficult even for the two inspectors to speak with authority on the state of the recruits in their seven districts. They had to rely almost entirely on hearsay evidence about the efficiency of the 10,000 or more men for whom they were responsible.

We were indebted to Max Manus for a dispassionate attempt to discover the standing of *Milorg*, first in the eyes of the people generally, and second in the eyes of *Milorg* members. When he was in Oslo in the spring of 1943 on Operation *Mardonius* (see Chapter 5), he conducted a 'Gallup poll' on the subject, which in the nature of things had to be based on a small sample—to conduct any poll under the noses of the occupying power was a daring feat. His conclusions may have been rather more subjective than those of the latter-day poll. He questioned forty men in the street who had no connection with an underground organization, thirty-five of whom had 'a more or less thorough knowledge of *Milorg*, and usually represented one or another degree of scepticism and aloofness'. Five said they knew Lieutenant-Colonel Ole Berg had been one of the leaders—as did all the members of *Milorg* whom Manus met, a reflection on the security of the organization. The older people he questioned said: 'What can our untrained youth achieve against German elite troops?'; 'London says we must not start things which may cost too many lives'; 'If you get mixed up with *Milorg* you're sure to be arrested.' Out of twenty younger people, eight had been members of *Milorg* in the beginning, but had resigned 'because nothing had come of the work'. The Germans knew all about the secret army, and would completely break it up when the invasion came. One man said it was obviously in the German interest that Norway's best youth should remain 'nicely and prettily quiet, working flat out for the Germans all day, and holding fearfully secret meetings in the evenings'.

So much for the views of non-members. Manus took statements from eight members of *Milorg*, one section leader, and seven privates from different sections of the organization. Of the latter, he classed five as military analphabets. Their knowledge of guerrilla technique, sabotage, and warfare generally was 'only what an interested youth can pick up from newspaper reading, films, discussions with his friends, and his own imagination'. Three had some knowledge of explosives and fuses (including plastic explosive and cordtex) and thought they could blow up a railway line. One even thought he could manage a bridge. 'We taught the section leader how to handle our arms: he was very interested and asked if he could keep for his section those which we did not need ourselves.' Every one of the eight began and ended every discussion by criticizing the leadership. They were not competent to tackle either current duties of organization or eventual duties in the field. Further, in Manus's opinion all had much too intimate a knowledge of *Milorg*'s command structure.

The section leader came in for high praise. He

> was decidedly one of the finest types we met at home. A simple sound patriot who fought entirely from a sense of duty and feeling for his country, free of many of the less glorious minor motives which so many of the rest of us possess. He was a man who saw it as obvious that he should help us with our job, without for one second thinking of pay or glory. He was the only one we met who was

prepared to accept on this basis the fairly certain risk of death we had to offer him.

Manus was critical of the universal acceptance of the slogan 'Wait for the invasion' which had, he believed, penetrated into the very blood of the members.

This is also due to the fact that waiting can be quite a pleasant form of contribution, whereas all other forms may carry with them danger and unpleasantness. The 'Wait for The Day' motto is fairly bombproof today because it suits people well, but we have a strong suspicion that on The Day we shall run a risk of people saying to themselves 'We'll wait for another day'.

Manus may have been right in criticizing *Milorg* for poor security; but the whole purpose of the secret army was to lie low until it was called into the field. A few may have regarded *Milorg* as a safe haven (although its poor security often made it anything but safe), but that does not justify criticism of the entire organization. However, it is easy to understand how one of Manus's temperament—who with a handful of kindred spirits was running incalculable risks in order to inflict immediate damage on the enemy—must have been irked by the undynamic *Milorg*.

SOE's agreement to work principally through the secret army leadership in Oslo tied its hands, as Munthe had suggested it must; and made it more difficult to play a part in implementing Allied policy in Norway. In the spring of 1943 a new difficulty arose. Four SOE missions—*Lark*, *Anvil*, *Cheese*, and Gunnar Sonsteby in Oslo—independently tackled headquarters about the value of the work they were doing. Wilson summed up their approaches. They were uncertain about their ultimate objectives, and whether the game was worth the candle. They saw that the risks in underground work were increasing and that the capture of agents and W/T operators in several areas had led to arrests and reprisals at a time when the probability of an Allied invasion was receding. Wilson had a good deal of sympathy with this approach. He himself was satisfied that the Norwegian high command believed there would be no invasion before a German capitulation in Norway; and that their sole interest was in providing cover for any interregnum at the end of hostilities. They were using SOE as a convenient means of policing the country until the German collapse, but 'we don't want to act as dogsbodies for the Norwegian government'. SOE's job was to attack the Germans, not defend the Norwegian government. The only reply he could give to the dissatisfied agents was that, so long as there was an underground movement, the size of which the enemy could not measure accurately, they would be compelled to keep very substantial forces in the country.

That it was sensible to subordinate SOE activities (other than *coups de main*, and operations in areas where *Milorg* was not represented) is witnessed by events on the ground when the organizations were working separately. In the Trondheim area, for example, the leaders of the large and old-established *Lark* organization saw the folly of competition, and took the law into their own hands. *Lark* became of its own freewill 'an accepted and integral part of the Secret Military Organization for the area'. Again, *Raven* (G. Merkesdal) in the Voss and Hardanger region, which had gone in to recruit guerrilla groups in July 1942 (when SOE were still trying to be independent) and which was joined by *Pheasant* (Bjarne Iversen) in March 1943—ostensibly because London thought *Milorg* was not well represented in that area—was well aware of the dangers of crossing wires. The leader warned London: 'As regards the military organization in Bergen we must come to an arrangement with them as the District Leader there reckons both Voss and Hardanger are in his area. There must be a certain amount of co-operation.' This was in March 1943, six months after SOE's decision to leave the secret army strictly in the hands of the central leadership. Nevertheless, Merkesdal continued to operate independently and in September 1943 reported that he now had 400 men, in eighteen groups with strengths varying from ten to fifty. His neighbour Iversen had more than 400. Merkesdal went out of his way to defend *Milorg*. He knew it had been criticized in London, where some were under the impression it was falling down on the job. This was quite untrue. 'We must repudiate the accusation and state that SL and his men have our full confidence.' Both Iversen and Merkesdal were co-operating with *Milorg* 'with full understanding'.

The further from Oslo, the more acute the problem of relationship. In 1944 Knut Aarsaether, in charge of SOE's *Antrum* organization in Alesund, was well aware of the anomaly of having two similar bodies rubbing shoulders, and indeed rubbing each other up the wrong way. He had no doubt which was the senior body. 'We have not made use of the existing *Milorg* people because there are so many of them who cannot keep their mouths shut, but we shall take a number of them later.' *Antrum*, training its own men in the use of Bren and Sten guns, would absorb *Milorg*, not the other way round.

The Norwegian Section continued to wrestle with the problem of maintaining its influence in Norway, while at the same time sticking to the agreement that *Milorg* should be allowed to make the running, with SOE as a sort of ancillary service providing arms, explosives, and instructors. In May 1943 SOE was still arguing that *Milorg* was an unknown quantity, in spite of its leaders' visits to Britain. It was vulnerable because it was having to face up to the threat of deportations to Germany, the enforced movement of people from one part of the country to another, which could play havoc with local organizations, the taking of hostages, and the execution of leading personalities. It was accepted that co-operation would become increasingly necessary, but this did not mean the amalgamation of SOE and *Milorg*. The maximum help must, of course, be provided for *Milorg* groups, but 'to achieve any results in this direction SOE must keep a presence in the field, and a certain amount of independence'. Since the Norwegian Section had no interest in groups 'outside SOE interests' (presumably this meant *Milorg* groups), it followed that FO IV could have no claim to intelligence about SOE organizations operating in Norway independently of *Milorg*. In spite of this specious reasoning, there was a new directive shortly afterwards stating that all special operations in Norway would be carried out under joint agreement between the Norwegian high command and SOE, using personnel trained by the latter.

In May 1943 representatives of FO IV, led by Oen and Schive, met representatives of *Milorg*, led by their commander, Jens Christian Hauge, in Stockholm, to clarify their relationship with each other and to discuss future policy. SOE deliberately refrained from joining the discussions lest it might seem to be committed to some line which would later prove incompatible with Allied strategy. A statement was agreed at the end of the meetings, the salient points being: Young Norwegians should continue to be trained in the use of modern weapons, which would release their urge for patriotic activity while waiting to take on the enemy. It would be up to FO IV to decide about the deployment of *Milorg* forces; and freedom of action given to districts in certain circumstances under SL's first directive was cancelled. W/T communication with Britain would be provided for all districts; FO IV would continue to send instructors to Norway to train men. *Milorg* was aware that FO IV was in touch with non-*Milorg* groups, and that similar groups might become necessary in districts where the *Milorg* organization had been broken; but at some stage these groups must be linked with *Milorg*. *Milorg* was aware that FO IV was co-operating in certain activities outside the sphere of the secret army, which for reasons of security must remain independent. Finally, there were some extremist groups in Norway which refused to subordinate themselves to *Milorg*. An effort must be made to keep Norwegian youth away from them.

Commenting on the statement, Wilson noted that the extremist groups were Communist, too few and too small to be of any importance; and he decided with satisfaction that the reference to activities outside the sphere of the secret army would allow the Norwegian Section to carry on with their own activities in Norway. Nielsen, head of the Norwegian Section in Stockholm, also had something to say about the statement. He took it to mean that the organization and training of *Milorg* were seen more 'as a safety valve for the potential hotheads than as a serious active contribution to the expulsion of the Germans'. He seems to have thought that *Milorg* should be making itself ready for a unilateral attack on the occupying forces in advance of an Allied invasion—which no one had ever suggested. This idea is strengthened by his other comment: 'Both the Norwegian government and the Secret Military Organization are more concerned with schemes for coping with the situation which arises when the Germans leave or are pushed out, than with any activity now.'

The suspicion that *Milorg*'s and SOE's objectives did not coincide exactly and that the secret army might not be the effective force the leadership claimed persisted in the early months of 1944. The Norwegian Section's confidence in the organization was not increased by a long statement from a leader who escaped to Sweden, and whose word SOE Stockholm entirely trusted, dismissing the idea he had a chip on his shoulder. He said the original intention was that *Milorg* should be a national secret army supplied with arms from Britain in order to back up an invasion, but 'during the war the organization has become obsolete'. Further, the Gestapo knew all about it, and had succeeded in wiping out many groups. Training, often carried out by partly qualified instructors and without arms, had not been effective. While the leaders had the best of intentions, it was deplorable that they would not admit that in reality the whole organization had no practical value. Reports like this were hardly calculated to encourage SOE to sit back and leave all activity in Norway to the secret army.

In March Colonel Oen paid another visit to Sweden to confer with the *Milorg* leadership; and again SOE decided not to take part in the discussions lest it be committed to some undesirable line. What had been clear to many for a very long time—that there would be no Allied military intervention in Norway—was now finally accepted. Oen was briefed in London to say that the agreed policy was to increase the attacks on enemy shipping, which would be carried out from Britain. The only other support in connection with the D-Day landings would be the cutting of communications, especially the railways. It was considered that *Milorg* was not yet sufficiently well trained nor equipped to be of any military value. The greater part of the discussions was devoted to the consideration of the different sets of conditions in which Norway would be liberated, and the part *Milorg* would play. The need to take special precautions to protect industrial plants and key points ranked high on the agenda; and Hauge laid great stress on the need for *Milorg* 'to initiate certain local actions' against the occupying power, both to hit back at the Gestapo and to keep the resistance movement on its toes.

The relationship between SOE groups still operating in *Milorg* territory again came up for discussion. The *Pheasant* group was now stronger than ever, having 1,000 men; and one of the decisions of the conference was that the group should be allowed a free hand, with authority to override arrangements made in the area by *Milorg*. The central leadership undertook to send messages making this clear to the local leader. The plan worked reasonably well except for a problem posed by the former *Milorg* leader, General Munthe. He was nearly 80 and very deaf, and his group 'had taken matters out of his hands without ever telling him they had done so'. The General was, therefore, still under the impression that he was in command, and he could not understand *Pheasant*'s place in the scheme of things. He was sent to Oslo so that the position might be explained to him, but it made no difference. For a time he said he would take his orders only from General Hansteen in Britain, but on reflection he decided that Hansteen could not possibly know what the local situation demanded. The solution was found when Iversen agreed with the other local leaders that they would all pretend to be under Munthe's command 'just to keep him quiet, but to take no orders from him, and not to tell him anything'.

It is uncertain what contribution Norway's secret army would have made had it been called on to support an Allied invasion in the early years. The returns made from the field were not encouraging. They show that in many districts the leadership had suffered from the attentions of the Gestapo, and also that the true state of affairs in many districts was often unknown. In District 11 the number of men who could be called on could not be ascertained, but it was admitted their training was behindhand. In District 12 'the number of men cannot be stated'. Nor can an estimate be given for Districts 14, 15, 16, 17 and 18: 'we have made repeated attempts to get in touch with possible groups without success.' In District 20 contact had always been slight, and the Oslo leadership had no real contact with Districts 21 and 22 in the Trondheim area. They had sent money and ration cards to a group believed to be in difficulty there, but they had no idea who they were.

Milorg's leadership faced the problem which resistance leaders had to face in all occupied countries. How to equip and train an army in secret, how to keep morale at a high level when no action was possible, how to ensure that, when it was called to open conflict, it was as near the peak of operational perfection as possible. Until the end of 1943 the policy of the *Milorg* leaders was to avoid sabotage—which would lead only to savage reprisals, perhaps doing more harm than good—and to concentrate on training. Nevertheless, many of the patriots of *Milorg* contrived to play an active role during the earlier period when the order of the day was to lie low by helping sabotage operations mounted by SOE.

> Even if sabotage operations were originating from Great Britain, and even if the agents were under directives to refrain from involving *Milorg*, it could not be avoided that *Milorg* people got involved . . . I doubt whether you can find any SOE sabotage action not being assisted by *Milorg* members in one way or the other, even during the period when they were not supposed to do so.[4]

It is impossible to do justice here to the help given to the Allied cause by all these members of *Milorg*. Perhaps one may be allowed to represent his fellows. Olav Skogen, founder of *Milorg* in the Rjukan area, was arrested in December 1942, and, although tortured by the Gestapo, disclosed none of the important information in his possession. His services later proved of great value in the supply of intelligence and the transport of stores in connection with the crucial attack on the heavy water plant at Vemork in February 1943. He was but one of many *Milorg* men who played a part in the sabotage campaign, without prejudicing the security of the secret army preparing for The Day.

9
Coup-de-main operations to D-Day: Norway

THE sabotage sponsored by SOE in Norway before the middle of 1942 was no more that isolated pinpricks. It enhanced the image of an amateurish and incompetent organization created by Rickman. A shining exception was the success of the *Lady* expedition, which had virtually no training in Britain and no pre-arranged reception in Norway (see Chapter 2). It may have suggested that the escalation of sabotage into a major weapon was a simple matter, and that, given the resources and willing volunteers, there was no limit to what could be achieved; but alas, this notion was swiftly disproved.

Railway sabotage

Operation *Barbara*

In January 1941 Malcolm Munthe told London he proposed to send a team to sabotage the railway from Trondheim to Storlien just over the Swedish border, where he thought there would be a good chance of success. He asked if there were any explosives dumps in the area which his men could use. It was explained that all earlier dumps were almost certainly known to the Germans and it would be suicidal to try to draw on them. A few days later he signalled London that rumours about his sabotage plan were already circulating in Stockholm and that the police were interesting themselves in the proposed operation. He must therefore lie low for a few days—surely a wholly inadequate precaution, since there was not the slightest hope that the police, under orders to weed out anything that looked like espionage or sabotage based on Swedish soil, would in these few days close their file on the subject.

Almost immediately Munthe said he was sending 'a gang' to reconnoitre the railway and then find a place on the west coast where the requisite explosives could be conveniently landed. The gang would then go on to Shetland for a brief course on demolition, leaving two of its number behind to provide a reception committee for its return. London liked the plan. A boat would come to pick up the group—known first as

Freddy from the name of the leader, Arne Frederiksen, and later as Operation *Barbara*—within five days of an all clear signal from Munthe, given reasonably good weather. Munthe assured London that there was nothing to be worried about, despite the unhealthy interest of the Swedish police. The members of the *Freddy* gang were all under the impression that they were working for *The Horse*, the fictitious agent he had created (see Chapter 3), and Munthe guaranteed that they were totally unaware of his connection with the British Legation. To consolidate the image of *The Horse*, he had sent them good luck messages typed on serviettes, the paper that best reproduced his red ink rubber stamp of a prancing horse.

The gang left for Norway on 6 February 1941, being escorted to the border by William Millar, working for SOE as a cut-out. Each man was provided with his own rations, a compass, a sleeping bag, and Kr. 600. All went well, except that one man became detached from the party in a thick fog, and failed to cross the border. The others went on to reconnoitre the railway as planned, and then looked for a landing place for the Shetland Bus that would collect them. They would mark it with a clothes line on which would hang 'three red cloths', a form of signal later used by SOE in the Far Eastern theatre, which seems somewhat inappropriate. Ideally the spot chosen would be far from human habitation, and the casual observer would no doubt ask himself why anyone should hang out washing there. This point seems to have occurred to *Barbara*, for in the event the signal was not used.

Munthe told London on 1 March that the team was now waiting 9 miles south of Kristiansund, and the Shetland Bus—the *Mars*—duly arrived there on the morning of 5 March. She passed close by two German patrol ships, which paid no attention, and later met three more and a 3,000-ton cargo vessel escorted by seven small armed trawlers. Having searched in vain for the three red cloths, the members of the team put into Kvisvik harbour to consider their next move, but, when people took an interest in the visitor, they moved out to sea again. Finally, they took the bull by the horns and returned to Kvisvik to moor alongside the jetty. To their great relief Frederiksen, who had been watching their manœuvres and had divined they were the Shetland Bus, was now waiting for them. He explained that the rest of his party would meet them at sea, so they left quickly and had no difficulty in making the rendezvous 7 miles out. The *Mars* returned safely to Lerwick on 8 March, carrying all eight men, as the plan to leave two as a reception committee had been abandoned.

Meantime hasty preparations were being made to instruct the group in the art of railway sabotage. A special consignment of rails, fishplates, spikes, and sleepers was brought from Aberdeen, so that practical training could be given. An instructor nominated by Gubbins had mixed feelings about his pupils. He was astonished to find that they had not the slightest idea what was required of them. None had ever handled explosives, and he thought the wrong man had been appointed leader—simply because it was he who had made the first contact with the Legation. The instructor thought Stockholm should have given the men a clearer picture of what they were in for. They should have been told that, unless they could achieve 'prolonged disorganization', they were wasting their time, running great risks for a negligible reward. They had failed in their preliminary reconnaissance to identify worthwhile targets, and had been thinking in terms of bridges, which would call for much greater quantities of explosives than they could carry and would take many hours to prepare for demolition. On the other hand the six Norwegians and two Swedes were a very fine bunch of men, exceptionally quick in picking up the essential points in explosives work during their week's training, and showing no sign of fear in handling the charges. Their specialized instruction concluded with blindfold practice, so that they would be able to operate in the dark. There was a contretemps at the end of the brief course when the explosives outfits ordered from Station XII were found to be deficient. There was little time to get replacements as it was essential to get the men back to Norway before a new system of identity cards made their forgeries out of date. Eventually, after some angry telegrams, each of the six men detailed for the railway sabotage was satisfactorily equipped 'as a self-contained unit'. The task of the other two was to set up a large dump of explosives for future use, and then return to Shetland.

There was a good deal of alarm and despondency when it was realized at headquarters that one of the original party had become detached in Norway (he had been replaced by the leader's brother) since it was conceivable that he would reveal the whole enterprise, possibly under duress. The whole *Barbara* team was interrogated at great length about its activities in Norway in general, and the disappearance of the colleague in particular. The Norwegian farmer who had given them shelter on their first night had talked about the severe punishment the Germans meted out to people caught with firearms, and it was concluded by his companions that the missing man, who alone among them had had a revolver, had got cold feet, and when the fog came down saw his chance to escape. The members of *Barbara* were unanimous that he would not talk, and it was therefore agreed that their mission should go ahead.

The party left Lerwick on its return trip on 7 April (after a false start due to engine trouble, and a second trip when it failed to find a suitable landing place) and disembarked safely at Klungervik on 10 April. They hid in a nearby farm and then during the night of 11–12 April set off for their objectives, which had been chosen after discussion in the Shetlands. The pair responsible for setting up the explosives dump successfully accomplished its task and returned to Lerwick on the *Vita*—which had replaced the *Mars* when her crew refused to undertake any more ferry services. The other three pairs were supposed to make three separate attacks on tunnels on the Trondheim–Storlien railway between Opdal and Berak, between Storen and Kotsoy, and at Gudaa. In the event only the last was carried out. On the night of 17–18 April Frans Hellstrom and Gudmund Nygaard placed an explosive charge to the east of Gudaa station which would go off when the next train went through the tunnel. Unfortunately the next train was a railcar carrying a few Norwegian passengers. Some windows were broken and five people were cut by flying glass, not an impressive dividend. The other pairs were deterred by men working on the stretches of line assigned to them, and all six made their way back to Sweden, where they were arrested and charged so quickly that it seems the police were lying in wait. William Millar, the cut-out Munthe had been using to organize the *Barbara* party, had already been arrested—further confirmation that the Swedish authorities were aware of what was going on. Millar and *Barbara*'s leader were sentenced to eighteen months, the others including the Swedes to terms ranging from three months to a year.

The inquest into the fiasco was conducted in London and Stockholm. SOE told Munthe that the projects envisaged by the gang before they left Norway were quite impracticable. In future it would be much better to work the other way round. London would select the target, which would be examined on the ground before an operation was mounted. It could be photographed and models made for the use of teams being instructed in Britain. Another alleged weakness was that all eight men were dressed so alike that they seemed to be in uniform. This would not matter in the remote areas but it seemed likely to arouse interest in towns and villages. In future large groups should be asked to dress differently. (Perhaps a curious suggestion. Probably Norwegian civilian outfits tended to be uniform, and to insist that a large party dressed differently might also arouse interest.)

The Orkla mines

Operations like *Barbara* did little for SOE's image. By contrast, a series of attacks on the Orkla pyrites mines in northern Norway starting a year later when SOE was getting into its stride showed that, with a sensible choice of target and careful planning, special operations could weaken the enemy war effort. The Germans were drawing pyrites from the Orkla mines for the manufacture of sulphuric acid, a key chemical for many industrial processes. Although the quantities involved—50,000 tons a month—were relatively small, every little helps, and if supplies could be cut off it would be a case of every little hinders, and the destruction of the mine would boost the morale of the resistance.

Operation *Redshank*

Peter Deinboll, son of the engineer of the Orkla company and a member of the Linge Company, knew the area well and volunteered with two companions, Per Getz and Thorlief Grong to attempt to sabotage the plant (Operation *Redshank*).* They left Lerwick on 17 April 1942 on the *Harald* and when they reached the Halten fishing bank transferred to a fishing boat whose skipper promised to land them at the nearest convenient point, in spite of having to run the gauntlet of several German watch boats. Their plan was to destroy the generators and transformers of the Bardshaug converter station, an integral part of the Lokken power grid which supplied the mine and the electric railway which carried the ore to Thamshavn for shipment. The three had no difficulty in overpowering the small guard and laying their charges, and as soon as the work was complete Getz and Grong started their trek to the Swedish border, where their cover story was to be that they were wanted by the Germans in Norway for distributing illegal newspapers. Deinboll stayed behind to witness the results—which exceeded his expectations. He was only a few hundred yards away when the explosion occurred. The fact that all the windows of the building housing the electrical plant were heavily boarded intensified the effect of the explosion as the walls collapsed and the roof was blown skywards, 'smashed into little bits'. They had included some incendiaries which started a fire that burned for hours. Local people later reported that the

* All three members of *Redshank* later lost their lives on active service. Getz and Grong were lost with the submarine *Uredd* (presumed mined) *en route* for an operation against the Sulitjelma mines in Nordland in February 1943; and Deinboll in November 1944 in an aircraft bound for the Oslo area which was never heard of again.

mines were brought to a standstill, and that because of the lack of electric power the railway had to change over to steam. The Orkla Metal Company which depended on the mines for raw material for the manufacture of sulphur was also brought to a standstill. All three men got safely across the border into Sweden where their cover story was accepted. In Stockholm they first went to the Norwegian Legation where the Naval Attaché handed them on to Croft in the British Legation to arrange their transport back to Britain.

Operation *Granard*

Deinboll planned a second operation (*Granard*) against Orkla at the end of 1942. This time his companions were Sergeants Pedersen and Saettem, and things went rather less smoothly. Their objective was to sink an ore ship and to burn the loading tower on the jetty. They left Lerwick on 6 December 1942 in the *Aksel* (which was lost without trace on the return trip) with 5 cwt. of stores. They had to take to their motor dinghy 18 miles from their destination because *Aksel*'s skipper feared that sparks from her funnel would be spotted from the shore and after a very difficult trip, thanks to their heavy load, rough seas, and a compass affected by the magnets of their limpet mines, reached the tiny island of Gronningen. When they launched their boat next day it promptly sank, but they managed to rescue their stores. The following day two boys came from the neighbouring island of Grip to collect driftwood. Deinboll and his companions, who were wearing sweaters over their British uniforms, extracted a promise from them that they would keep their mouths shut and started off rowing the boat, the engine now being completely useless. The boat took water again, and they were only just able to make it back to the beach, and salvage their equipment for a second time. They now asked the two boys to take them to Grip where, when no one was prepared to risk putting them up, they were allowed to spend the night in the village hall. Fortunately the German guard of six men was stationed on another island, 300 yards away, and since they only had a rowing boat there would be plenty of time to see them coming.

Deinboll now found a man who agreed to swop his rowing boat for their motor dinghy, but the rowing boat also turned out to be useless. Two young men 'whose spirits had been greatly strengthened by the help of one of our large bottles of rum' were now called on to fetch their equipment from Gronningen, but the rum proved too much for them 'and we were shipwrecked for the third time'. At the end of this episode the trio found themselves marooned on the tiny island of Ingripen, 2 miles east of Grip, having sunk the useless motor dinghy, and been promised by the two befuddled young men that they would come back with a motor boat to take them to their original destination. Four days passed without any sign of the motor boat while Deinboll and his companions sheltered in a hut intended for fishermen surprised by bad weather, using the furniture to keep the stove going in the absence of any other fuel.

As they clearly could not remain there indefinitely, they constructed a raft out of the hut door, some empty petrol tins, and driftwood. Deinboll was about to set off on this precarious craft, hoping to reach Grip where he would steal a boat and return for the others, when two more boys arrived in search of driftwood. They decided to hold them until dark and then use their boat to escape; but before they could leave the boys' elder brother, alarmed by their long absence, came to see what had happened to them. When he heard their plight he said he would return with a motor boat at midnight. At the risk of being marooned for another spell, Deinboll allowed all three brothers to return to Grip; and this time his faith was justified. The boy returned just before midnight, they collected their equipment, and were duly delivered at their destination on the mainland. They had been shuttled round the islands of Gronningen, Grip and Ingripen for exactly a week, and it must have been a great relief to know they were moving forward again.

On 20 December they made contact with Deinboll's father and were able to get the latest information about the security arrangements at the jetty. On Christmas Eve word came that the Germans were making a house-to-house search which presented no problem since Deinboll senior 'had arranged in his house a most ingenious hide-out, intended originally for black market supplies' in which there was just enough room for the three visitors. It was now decided that they would concentrate on sinking a ship and abandon the idea of burning the loading tower as it seemed unlikely that it could easily be destroyed by fire. A hut was found where their equipment could be kept until a suitable ship tied up at the export pier, but it was not until 26 February that the 5,000-ton *Nordfahrt* presented herself. That evening the three set out from the hut, having previously cached the civilian clothing needed for their escape to Sweden. It took them three and a half hours to get to their advance base in conditions which made skiing very difficult and a further three-quarters of an hour to make their final preparations and get down to the quayside. They managed to cross the railway line and the main road unseen, in spite of bright moonlight. They found a rowing boat and approached their target with the limpets hanging over the side in a triangle, already submerged at the correct depth. They had no difficulty in attaching them, protected by the shadow on the seaward side of the ship, but the wind made the return trip very slow. It took them an hour to get back to their starting-point, where they left the boat exactly as they had found it. When they got to the place where they had dumped their civilian clothes, a message came that their contact who had been going to help them on their way had been arrested, and they had to improvise. They contrived to get a motor van from Trondheim which took them as far as Hegset, and left them to make the rest of the journey on skis.

According to Deinboll's account:

It had been our intention to make the Swedish frontier on the same day, but we were overtaken by a heavy snowstorm. It was decided to try to reach the nearest hut shown on the map, some three and half miles east of us. In order to get there we had unfortunately to climb a hill. The descent from this hill into the Gaastjern Valley was far steeper than I suspected. In fact, we realized later that there were precipices almost everywhere except at one point. Visibility was only about two yards and we had to keep close together. I tried to descend sideways very cautiously but suddenly slipped over the precipice. I fell a considerable distance, landing on a steep snow-covered slope at the bottom of the valley. During my fall I hit a ledge and injured my left ankle.

As he could not move, he dug himself in where he fell. A break in the weather allowed him to drag himself to a hut 300 yards away before the storm closed in again. His new refuge was half filled with snow, with the wind blowing right through it, but he had to stay there until 3 March with nothing more to survive on than snow melted in a cup held between his thighs. He knew he dare not remain there as he was beginning to suffer from frostbite, so he struggled on another 2 miles when he came across a very good hut 9 miles from the border, complete with stove, fuel, and a store of food. Here he rested until 6 March when he felt strong enough to complete the journey to the frontier.

Pedersen and Saettem had searched in vain for their missing leader, and they had failed to hear the pistol shots he fired to attract their attention. They got into Sweden without difficulty, and reported that Deinboll was missing. Their belief that he could not possibly have survived was disposed of when he eventually turned up in Stockholm, after having a difficult time with the border authorities who recognized him as the man who had passed through their hands in similar circumstances early the previous year.

It was established later that the attack on the *Nordfahrt* was less successful than had been hoped as it was possible to run her into shallow water after the explosion. Nevertheless, the ship was out of commission for many months.

Since it was very likely that Deinboll's father would be suspected of complicity in the attack, it had been agreed that he and his wife and daughter should escape to Sweden, and thence to Britain, which they successfully accomplished.

Operation *Granard*'s accounts show how meticulous SOE's men could be:

	Kr. Nor.
Deinboll	
Compensation for 2 housebreakings	510
Expenses in Sweden	227
Pedersen	
Clothes in Sweden	120
Suitcase	95
Hotel	25
Food	35
Train journey	40
Wrist watch	363
Saettem	
Clothes in Sweden	210
Suitcase	115
Hotel	40
Food	30
Train journey	35
Various	100
Joint account	
Help from the people of Grip	250
Motor boat from Ingripen	2,000
Accommodation, Ringdalshavna	30
Helpers (including 'our maid')	3,050
Horse and sledge	3,750
Food	250
Journeys to the town	400

Rent of hut	200
Petrol for car from Trondheim	50
Grand total	11,925

Granard's operational instructions included the provision: 'The party will avoid the payment of large rewards to loyal civilians for services rendered in simply a spirit of hospitality i.e. when no knowledge exists of the party's origin and tasks.'

Operations *Feather*

The capture of Sicily by the Allies reduced the quantities of sulphur available to the Germans, and supplies from Norway became relatively more important. In the autumn of 1943 it was therefore deemed expedient to mount another attack on Orkla (Operation *Feather I*). Deinboll was again the obvious choice for leader and on 10 October 1943 he and six others were dropped west of Orkdalen. The party landed without mishap—surprising in view of the unfavourable nature of the terrain—and on 31 October put out of commission five of the fourteen locomotives of the Lokken–Thamshavn railway, the only outlet for the ore, and in addition did considerable damage to the engine and wagon sheds. The party boldly remained in the area to carry out a second attack on 21 November, when a shunting engine was destroyed. Unhappily the second in command, Fenrik Odd Nilsen, stumbled and was killed by the charge he was carrying, and on the journey to Sweden a second member of the team, Private P. Skjaerpe, was captured and subjected to brutal treatment by the Germans. He survived, however, and was liberated at the end of the war.

Deinboll returned to Britain, but the rest of his party remained in Sweden with a view to carrying out yet another attack at Orkla, which they did in May 1944. This (Operation *Feather II*), the last *coup-de-main* operation planned from outside Norway, was necessary because production and export of pyrites was on the increase thanks to various makeshift measures to get round the difficulties caused by the earlier attacks. On 9 May 1944 Fenriks Heggstad, Bjornaas, and Wisloff stopped the last remaining heavy locomotive near Hongslo, took off the crew, and completely wrecked the engine. As their predecessors of *Feather I* had done, they remained in the neighbourhood, in spite of frantic enemy activity, and on 1 June destroyed a powerful rail bus in the same way. An attempt to wreck a bridge at Svorkmo was abandoned, and the whole party got out safely to Sweden.

Heavy water

If the series of attacks on the Orkla mines and their railway system impeded the German war effort, the attacks on the hydrogen electrolysis plant at Vemork in Telemark may have prevented Germany from winning the war. She relied on this plant for 'heavy water' (deuterium oxide) for her atomic bomb programme, and, while there is now some doubt on the point, it certainly seemed at the time that, given large supplies of heavy water, German scientists might win the race to the bomb, with what consequences it was then too awful to contemplate.

Operation *Swallow/Gunnerside*

The campaign against Vemork effectively began in March 1942 when Einar Skinnarland (*Grouse*), who had escaped from Norway with Starheim on the *Galtesund* (see Chapter 4), parachuted to the neighbourhood of the plant to transmit intelligence about activities there. In July 1942 the War Cabinet instructed Combined Operations to carry out an attack, and on 18 October an SOE party (*Swallow*) of four Norwegians led by Captain Jens Poulsson dropped into the nearby mountains—two earlier attempts having been frustrated by cloud—to prepare the way for the Combined Operations expedition. At that altitude and in bitter cold a man could carry no more than 65 lb.; and, since their equipment weighed a total of 520 lb., each man had to make three journeys every day over rugged ground and knee-deep snow—taking 65 lb. on the first trip and then going back for the second and third loads. Lakes and marshes were not fully frozen and their feet were permanently soaked. When, after fifteen killing days, they reached their forward base, Poulsson and Arne Kjelstrup went ahead to reconnoitre the plant while Claus Helberg returned to a deserted farm 'to steal all the food he could carry', which involved a round trip of 50 miles. On 3 November, just as their W/T operator (Knut Haugland) got through to London, the battery failed. They got a replacement from the keeper of a nearby dam, and were now at last ready to receive the Combined Operations party (*Freshman*), which left Wick in Scotland in two aircraft towing gliders. One aircraft and both gliders crashed 100 miles from Vemork. The survivors were captured and shot by the Germans in spite of the fact that they were in uniform.

Six men from the Linge Company trained to make a second attempt; but now the Germans knew the plant was threatened and strengthened its defences. The advance party waited patiently while their new battery gradually lost its power. Poulsson recorded: 'Everybody except myself went sick with fever and pains in the stomach. We were short of food and had to eat reindeer moss. I went out every day after reindeer but could find none. Our dry wood came to an end.' Then at last he bagged a reindeer 'which made for a happy Christmas celebration'.

At midnight on 16 February 1943 the six Linge men (code-name *Gunnerside*) dropped to a frozen lake 30 miles north of *Swallow*, whose signal lights had helped them to take their bearings. (The six were Lieutenant Joachim Ronneberg, Captain Knut Haukelid, and Fenriks Fredrik Kayser, Kasper Idland, Hans Storhaug, and Birger Stromsheim.) The flight had taken place in good weather with bright moonlight. The six jumped from 1,200 feet—'exit according to plan, and in perfect order. Highly successful, with men in good spirits.' Because of the size of the team the dispatchers had to get them out of the aircraft quickly, otherwise the great distance between the first and the last would make the assembly of the men and the collection of their equipment difficult. The dropping order was carefully worked out. First went six containers with brown parachutes as fast after each other as was humanly possible. Then three men with white parachutes, followed by a package containing their skis, and a second their rucksacks. Next, in the centre of the drop, came the toboggans for the whole party, and finally a repetition of the first gambit—the second three men, their skis, and their rucksacks. So well did the dispatchers do their job that no more than 800 yards separated the first and last parachute. The only problem was that one of the rucksack container's parachutes failed to collapse on the ground (always a possible danger—the agent could manipulate the harness of his parachute, but one carrying stores was at the mercy of the wind) and was blown nearly 2 miles at a speed no skier could match. Happily 'it found its way into the only ice crack we came across' and the telltale evidence was retrieved and buried in the snow along with the other parachutes. The only damage was to the frame of one rucksack, and to a sleeping bag which was torn. They had an early difficulty when they ran into a reindeer hunter (an occupational hazard) whom they had to keep under guard for several days. When he swore to keep his mouth shut they gave him money and food and sent him on his way hoping he would be as good as his word.

A week later, in the words of *Swallow*'s leader,

> we were alarmed to see two bearded civilian skiers in apparently first-class physical condition. I ordered one of my men to put on his camouflage ski smock

and a civilian ski cap. He set out to make contact with the strangers. If questioned he was to say he was a reindeer keeper on his rounds. The rest of us went into cover.

There followed a long and sinister silence, then above the noise of the wind came 'three wild yells of pleasure'. *Gunnerside* had joined forces with *Swallow*.

The plan was for a covering party to cut the plant's perimeter fence, and lie in wait to support a demolition party which would go inside the building. If anything went wrong, each man must use his own initiative to make the operation succeed. No torches would be carried; weapons would be unloaded to avoid the danger of an accidental shot. Anyone captured was pledged to take his own life. At 8 p.m. on 27 February the two parties left their base, first on skis, later on foot, sinking up to their waists in soft snow. They had to take cover when two buses carrying the night shift appeared and then began the steep and slippery descent to the river Maana, on which the turbines depended for their water supply, to find that the ice was breaking up and that the only means of crossing it was a precarious snow bridge with several inches of water sweeping over it. After scrambling up the precipitous rock face on the other side they rested and had a snack.

At 12.30 a.m. they approached the factory gates and had no difficulty in cutting the chain securing them. The covering party took up its position while the six-man demolition team crept towards the basement door of the five-storey concrete building giving access to the most vital part of the plant and the heavy water storage tanks, only to find the door locked. The only other way in was through a funnel for cables and piping which the leading pair found with some difficulty, and managed to negotiate. The solitary Norwegian guard offered no resistance as the pair placed the explosive charges. 'This went quickly and easily. The models on which we had practised in England were exact duplicates of the real plant. I had placed half the charges in position when there was a crash of broken glass behind me.' The other members of the demolition party, unaware that their leaders were already inside, had failed to find the tunnel, and decided to smash a window. After the charges had been double checked, they told the guard to get out, and themselves left rapidly. They had gone only a few yards when they heard 'a cataclysmic explosion', which drew the sentry from the main gates and left the way clear for their retreat.

Along with the key plant, 3,000 lb. of heavy water were destroyed.

Five of the *Gunnerside* party reached Sweden at the end of March 'in excellent spirits and condition' after a 250-mile journey on skis in vile weather, having endured almost unbearable hardship. The sixth, Haukelid, stayed on for another year. The *Swallow* party also remained, to report the effects of the operation. Helberg was pursued by three Germans in the Hardangervidda mountains for two hours before one got within striking distance. Helberg fired a single shot which missed, but induced his adversary to panic and empty his whole magazine ineffectually. Having thus shot his bolt the adversary turned tail, but 'I sent a bullet after him, he staggered and finally stopped, hanging over his ski sticks'. Almost immediately afterwards Helberg fell over a cliff, broke his right arm, and next morning walked into a German patrol. They accepted the story he told, and sent him to a hotel to await treatment. Reichskommissar Terboven and his entourage arrived the same day and displaced many guests, but, because of his injury, Helberg was left undisturbed. Unhappily a lady guest whom Terboven had selected to 'entertain him' refused to co-operate with the result that all the remaining guests, including Helberg, were 'bundled into a bus and sent off to the Grini concentration camp'. In spite of his broken arm Helberg jumped from the moving bus, evaded his pursuers and their pistol shots, and in due course found his way back to Britain.

As it fell out, the *Gunnerside* attack, and a later bombing attack by the USAAF, effectively ended the production of heavy water at Vemork, but there remained substantial stocks. When Skinnarland later reported that they were about to be removed to Germany, the War Cabinet asked SOE to destroy them *en route*. The Germans were now on their toes. Special troops had been drafted into the area, aircraft regularly patrolled the mountains, and guards were stationed along the railway line that would carry the heavy water on the first stage of its journey to Germany—to the ferry at Lake Tinnsjo. It seemed virtually impossible to get through these defences, but when invited to do so, Haukelid approached the problem with the same care and intelligence shown in planning the original sabotage of the plant.

He first considered the alternative methods of attack. The heavy water flasks could be blown up at the plant before dispatch. Or they could be attacked on the train from Vemork to the Tinnsjo ferry. Or he could wait until they had been loaded on to the ferry and sink them with the ship when she sailed. Or he could attack the consignment on the next stage of its journey—by rail from the other end of Lake Tinnsjo, say just before the town of Notodden. Or it could be blown up on the quayside before it was loaded on to the steamer at Skien. Or the steamer could be limpeted and sunk on the final stage of the complicated journey to Hamburg. Before making his choice, Haukelid went into Rjukan to pick up the latest news and gossip, which he found discouraging. The Gestapo had arrived in force, and the local *Milorg* leaders, whom he had been hoping to consult, had been forced to make themselves scarce. The Germans had let it be known that they fully expected an attack on the heavy water in transit, and they had at their disposal large numbers of police and troops. Their spotter planes made movement in the snow-covered mountains hazardous.

Haukelid decided there were only two possible courses: attack the train before it reached Lake Tinnsjo; or sink the ferry after it had left; and it seemed that the railway was so well guarded that there was no option but to go for the ferry. In case he failed, a party was told to stand by at Skien to make a second attempt as the heavy water was loaded on to the ocean-going steamer. Miraculously, in spite of the massive security elsewhere, 'not a single German guard had been posted on the *Hydro*'—the Lake Tinnsjo ferry. Haukelid, with Rolf Sorli (who earlier had provided invaluable information about the arrangements at Vemork, and had helped to plan the attack on the ferry) and another assistant, simply walked on board, to find most of the crew playing 'a noisy game of poker'. The engineer and stoker were working in the engine room, so there was no question of placing the charges there. In the passenger cabin they were accosted by the only guard—'thank God he was a good Norwegian!' He believed their story that they were on the run from the Gestapo, and gave them sanctuary. When they were alone, Haukelid wriggled through a hole and crept along the bilges to the bows, where he laid the charges, linked to two alarm clocks which he had adapted to detonate the explosives at a time when the ferry would have reached deep water. The trio left unchallenged at 4 a.m. and made their way to Oslo.

Seven hours later, when the *Hydro* had reached the deepest part of the lake, an explosion ripped a great hole in her bows, and, as she tilted forward, the railway wagons on her deck carrying the heavy water broke from their moorings and preceded her into the depths of the lake. The last remaining stocks of the chemical had been disposed of. Of the thirty-eight on board, mostly Norwegians, eighteen were drowned, including two brothers aged 14 and 15.

Power station sabotage

Operation *Knotgrass/Unicorn*

In September 1942 ten men from Combined Operations led by Captain Black and Captain Houghton, and two from SOE (Corporals Sverre Granlund and Djupdret) went by submarine to Bjerangfjord on the Norwegian west coast to attack the Glomfjord power station (Operation *Knotgrass/Unicorn*). The submarine surfaced 4 miles from the head of the fjord, having lain on the bottom a long time waiting for darkness to fall, and the party rowed the rest of the way. They hid the rubber boat and lifebelts and walked up the valley for about 3 miles, when Houghton and Granlund went ahead to reconnoitre the next stages of the route. According to Granlund's account: 'We passed a farm. During this stage I was rather afraid that the farm dogs would notice us. I found out later that all farmers had had to get rid of their dogs, as naturally enough, dogs must eat.' They found that the stretch of Svatrisen (the Black Glacier) which they had to cross was in good condition, with only a few crevasses which they could cross with the help of ropes; and next morning they set off up the steep mountainside and across the glacier. They had a near casualty when one man, very thirsty after the climb, tried to get a drink from a 'lochan' and fell into the water. Fortunately he managed to get himself out. They had reached the power station by the following day, when Granlund again went forward to make a reconnaissance. On 20 September the whole party was assembled at a point from which the attack could be made; and that evening the plan of campaign was discussed, especially the escape route to be used. There were two possibilities: one which meant an immediate climb of nearly 3,000 feet, which would be very exhausting, but had the advantage that no tracks would be left and there would be no need to cross the river serving the power station; the other was much easier but meant crossing the river twice, and leaving tracks in the muddy paths. Granlund favoured the stiffer of the routes, but the majority opted for the easier—a fatal mistake as it turned out.

A serious difficulty was that the mountainside went sheer down into a lake beneath, and it was almost impossible to avoid kicking stones which rolled right down to the bottom and made an alarming noise. They got down unobserved, and three men went to deal with the pipelines. The rest of the party went to the power station which they entered easily through a temporary wooden partition. They took the three Norwegian guards by surprise and were assured by them that there were no Germans present. They immediately walked into a German soldier, whom Granlund shot dead. The Norwegians then said the only other German had probably gone to Glomfjord, but there was a family living on the premises. The family was warned to get out, along with the guards, while the charges were laid, with 10-minute delays. As soon as the party was clear 'there was a dreadful explosion, followed by a second when the pipeline went up'.

It remained for the party to make its escape; and it now became apparent that they had made the wrong choice. They could not find the first bridge across the river and after much delay Granlund plunged into the very swift waters and swam across with the greatest difficulty. He had told the others to follow, but they did not, having seen how difficult the crossing was. He then made his way to Sweden on his own, and three of the others also managed to escape. The other eight were all captured, Djupdret, the other SOE man, having been fatally wounded. He was only 22 'and felt he was too young to die, but he consoled himself by saying that if a nation was going to live some must be willing to die'.

As a result of the attack the power station was out of action for many months. The pipes had been severed 350 yards up the hillside, and for eleven hours the water poured down, the mechanism which was supposed to cut off the water supply having failed. Forty thousand cubic metres of stones and earth were swept into the factory buildings. Two of the three turbines were completely destroyed, but the charge on the third had failed to explode. However, it was seriously damaged by the cascading stones.

The other main objective in Norway, apart from shipping, was the railway system. The possibility of interfering with German military traffic was constantly in the minds of the planners at SOE headquarters but in the early years they met with little success. The *Archer* and *Heron* parties sent into Nordland in 1941 and 1942 were given the task of reconnoitring the railway from Grong to Mo-I-Rana so that it could be effectively attacked when the time was ripe. The *Falcon* party was given the same objective early in 1944. All these parties met with disaster. As part of the support for the Allied invasion of the Continent, it was planned to cover the two main lines south from Trondheim, the interruption of which would hinder the dispatch of reinforcements from Norway. In October 1943 the Allied high command asked that plans should be prepared for the dislocation of railway communication in

Norway, but that no positive action should be taken in the meantime. Four specially trained parties were sent into the field: *Grebe* to Sollia in District 23 and *Lapwing* to Roros in District 22 in October 1943; *Fieldfare* to Romsdalen in District 21 in March 1944; and *Woodpecker* to Kongsvoll in District 22 in April 1944. Their parties were to reconnoitre the length of railway to which they were assigned, and also other local communications; to plan eventual attacks; to form dumps of explosives at strategic points near the line; and to recruit and train local men. They were not to attack without orders from London.

They were no more fortunate than their predecessors. Of the six men in the *Grebe* party, three dropped in a lightly-frozen lake and were drowned. The others withdrew to Sweden, returning to Norway in March 1944. *Lapwing*'s W/T was damaged on landing and it was not until June 1944 that the party was in touch with London. *Woodpecker*'s solitary agent was on his own and could do little before three others reached him in October 1944. The three men in *Fieldfare* established themselves in the mountains above the Romsdal Valley, but much of their food landed in a lake, and it was all they could do to survive. The failure of these missions mattered less since before D-Day SHAEF did not ask for large-scale railway sabotage (see Chapter 11, Norway).

10
Propaganda

SECTION D saw itself as both producer and disseminator of propaganda; but when the Rickman group disintegrated in 1940 the embryonic propaganda organization in Scandinavia went with it. It was decided, however, that, so long as Sweden remained neutral, Stockholm would be the ideal centre for spreading overt and subversive propaganda in Norway and Denmark, and possibly Germany and the Baltic States. Full use must be made of 'the sacred organization' of the British Legation there. The commercial and press departments should be asked to help, and also the 'lower hierarchy, the electrical, engineering and messenger staff'. By keeping the organization all British, it would be easier to get round the Swedish authorities, especially the Swedish secret police, who were controlled 'by a violently pro-German officer'. As Swedish printers would be reluctant to produce propaganda material for Section D, especially after the publicity given to the Rickman affair, it must have its own printing establishment in Stockholm to work on pamphlets and leaflets for Norway and Denmark.

It might be difficult to smuggle large quantities of propaganda leaflets across the Swedish border, and to distribute them in sparsely populated areas, but something could be done through chain letters and 'whispers'—rumours spread by agents. Leaflets should be dropped by the RAF only round about Oslo, where more than half a million people lived. It would be a waste of effort to drop them anywhere else. Lines of communication into Denmark should be organized through Stockholm, and via the consulates at Malmö and Hälsingborg.

There was no need to disseminate subversive propaganda in Sweden, except against individuals known to be pro-Nazi. It would be enough to make available Swedish translations of British pamphlets, although even that might fall foul of the severe anti-propaganda measure recently introduced by the Swedish government 'to prevent a foreign power from preparing intervention in Swedish affairs by carrying on propaganda or by otherwise endeavouring to influence public opinion'. Anyone offered money to distribute foreign propaganda, who did not report the fact was liable to imprisonment. Nevertheless, Section D thought British propaganda might even be welcomed. 'The Swede likes facts, quick repartee, and to be amused.' Since it was on the cards that Sweden might be attacked by Germany, or unduly influenced by her, it was necessary to have a propaganda organization that could operate under these conditions.

That was the future of propaganda in Scandinavia as seen by Section D; but plans had to be modified because of successive reorganizations of the British propaganda service. First Section D's functions were divided between SO 1, which became responsible for subversive propaganda, and SO 2, which carried on the rest of Section D's activities. Then SO 1 became part of the new PWE set up in August 1941 to provide propaganda against the enemy. This left SO 2 with only the distribution of propaganda in the field, reporting on its impact, and advising PWE on the conduct of its campaigns.

Although the operation of propaganda agents was not provided for in its charter, it was agreed that SOE, because of its facilities for training and dispatch, should provide instruction for them, initially at Station XVII at Brickenondbury in Hertfordshire (later known as Special Training School XVII) where two of the instructors were Kim Philby and Guy Burgess. By September 1941 it became apparent that two types of training were called for: specialist, for those who were to concentrate on propaganda, and elementary, for SOE's own agents, simply to make them more useful all-rounders. The specialist courses called for close collaboration with PWE's Country Sections, and a school for this purpose was provided near PWE's black headquarters at Woburn Abbey. It started as a civilian establishment under Captain J. W. Hackett, but in July 1943 became a war establishment with military personnel, and was transferred to Aldenham, Herts, as Special Training School XXXIX. The subjects studied were printing, propaganda writing and directives, and opinion research.

The division of responsibility between SO 1 and SO 2 led to friction. Rex Leeper of SO 1, in charge of the preparation of subversive propaganda, insisted on the right to send agents even to countries where SOE already had men in the field including Norway, Sweden, Denmark, and Finland. SO 2 were unhappy about this, since the activities of independent propaganda agents might cut across their own plans. They also suspected that Leeper was trying his hand at empire building. A *modus vivendi* was worked out. SO 1 would issue general instructions to the field, to go through the head of the SOE mission overseas, who could ask for them to be reviewed. A weekly meeting in London between SO 1 and SO 2 would keep each informed about the other's activities. SO 2 would continue to carry out the dangerous function of distribution, to act as a sounding board in the occupied countries, and provide intelligence on which propaganda themes could be based.

The relationship between the two organizations improved, although disagreement flared up from time to time. In July 1944 Wilson of SOE's Norway Section refused to allow the dissemination of material prepared by PWE specifically for a group of saboteurs trained in political warfare, their task being to subvert German troops. There was a storm of protest from PWE at this 'serious encroachment' on their work, but SOE were adamant. Sporborg said they had the right of veto; but in future the heads of the Country Sections concerned should try to settle their differences amicably.[1]

In Norway SOE had to contend with the Norwegian government in exile, which was jealous of its rights in the field of propaganda. It was early on authorized to control broadcasts in the BBC Norwegian service, except for straightforward news programmes; and it did what it could to influence the material devised by PWE for dissemination by SOE

> The Norwegian government regards all information propaganda and political warfare activities in Norway as being entirely its concern, as representative of the Norwegian people and servant of the Home Front. The government has always been and continues to be extremely jealous of what it considers to be 'interference from outside' in Norwegian domestic matters.

SOE found this attitude understandable, but somewhat exasperating. It seemed that the Norwegian authorities, like some of the Danish leaders, were more concerned with their own country's problems than those of the Allied nations as a whole.

That this made life difficult for the British side is illustrated by the fate of a scheme for operational propaganda—propaganda intended to influence resistance activities in the field, as distinct, for example, from propaganda to raise the morale of an occupied people, or lower that of the occupying troops. Or, to use SOE's formal definition, 'the use of persuasion and demonstration in an organized way to produce predetermined action'. The plan was that PWE would supply 'indoctrination and servicing in propaganda information, directives, material, etc.', while SOE would 'organize and operate the agents in the field and be responsible for transport, security, communications, etc.'. If this had come off it would have been the biggest single co-operative action by

SOE and PWE. That it did not was due entirely to the attitude of the Norwegians.

Fifteen members of the Linge Company were trained in the joint SOE/PWE propaganda school. The first were ready to go into action in September 1942, when the Norwegian Minister of Foreign Affairs (Trygve Lie) ruled that all internal propaganda must be the exclusive responsibility of his government. When it was agreed after much discussion that the propaganda groups should carry on, it was only on the understanding that the Norwegian Information Office in London would be responsible for the content of the propaganda, and that the dispatch of the agents should be handled by the Norwegian high command in consultation with SOE. Then there was a new complication. It was reported from Norway that the secret military organization considered that it was better qualified to conduct a propaganda campaign than anyone sent from Britain. It was therefore decided that, while SOE should continue to train agents, their services would simply be offered to *Milorg*. This was accepted by the Norwegian authorities, subject to *Milorg*'s approval.

Southern Norway was divided into ten propaganda districts. Three-man teams—leader, assistant, and W/T operator—would be sent to each district; and a representative of SOE headquarters would be posted to *Milorg* to ensure that the district organizations knew what was going on. This man (*Petrel*) left for Oslo at the end of November 1942 where he duly met the *Milorg* leadership. He reported from Oslo that the plan had been favourably received, and SOE was told that Trygve Lie would authorize them to go ahead. Before he did so, it was reported that, in spite of their original favourable reaction, a majority of the *Milorg* leaders now opposed the plan. Only a single instructor should be sent. In any case, it was added, morale was excellent, and it would be foolish to run risks for the sake of propaganda—although miniature radios and printing presses would be welcome. The Norwegians in London agreed. The increased strain on the resistance which the propaganda teams would impose would outweigh any benefits from their professional training. SOE found this disappointing, especially as it now had on its hands a large number of specialists with nowhere to go.

Radio had not been available to the propagandists in World War I, but now made it possible to reach the people of a country otherwise inaccessible. It allowed the British to conduct a campaign in Norway over the heads of the Norwegian authorities through a clandestine radio station (with the cover name of Research Unit (RU)) purporting to transmit within Norway, but actually situated in England, and run by SO 1. The RU could further Allied policies without necessarily having to seek the approval of the Norwegian authorities. It began to transmit in February 1941, and put out more than 600 programmes before it closed in December 1942.

It was some time before the Norwegians tumbled to the fact that the Freedom Station was British; but before then it crossed swords with SOE, which was working in close collaboration with the Norwegian authorities. Hambro protested in September 1941 that recent 'arrests and shootings and imprisonment of trade unionists in Oslo' were due to an ill-advised campaign by the RU inciting Norwegians to strike. He had always opposed this line'because it takes so very little to get them going, with the usual German reaction'. SOE was hoping that the RU would rather try to keep the Norwegians quiet so that they might be ready for the day when a serious operation would be called for. In October 1941 the RU was still a well-kept secret:

> Some people think the Norwegian Freedom Station is in a ship just off the Lofoten Islands. Others think it is in a lorry moving about somewhere in the western part of Norway. The important thing is that no one in Norway knows anything about us. One Norwegian with Nazi sympathies whose secret dealings with the Germans were exposed by the Freedom Station became so exasperated that he offered a reward of £5,000 to anyone who would disclose the whereabouts of the station . . .

The management of the RU recorded a list of the passive resistance measures they were urging on their listeners in Norway 'in all classes of society': telephone operators were urged always to give Germans a wrong number; waiters to give inefficient service to German officers ('there's a Colonel in today—don't forget to spit in his soup'); chambermaids to use only leaking hot water bottles if Germans were in the house. Presumably the propagandists took this sort of thing seriously—although it is difficult to believe that if every chambermaid in Norway had attacked every German officer with a leaking hot water bottle it would have had much effect on the morale of the occupying troops.[2]

Shortly after this the identity of the RU was revealed to the Norwegians. Trygve Lie summoned Tom Barman of PWE to show him a telegram from the *Chargé d'Affaires* in the Norwegian Legation in Stockholm summarizing recent transmissions attacking the Legation on the ground it was failing to further the Norwegian cause (which was generally agreed at the time). The *Chargé* guessed the station must be in England. Barman decided he must come clean with the Minister. He admitted that the broadcasts originated in Britain, and that the Norwegian government might have reason to feel aggrieved. In future he would keep the Minister informed, in broad outline, about the RU's plans. The Norwegian RU continued for another year, until it closed because of shortage of suitably qualified Norwegian staff, and the difficulty of getting news out of Norway, without which interesting programmes could not be made.[3]

The Danish RU fared better. In November 1940 Turnbull proposed a mobile Danish Freedom Radio station operating from a furniture van in southern Sweden, but there were several objections. By the time recordings made in Britain reached the transmitter—by air to Stockholm and thence courier—they would carry stale news. There were also technical difficulties. Further, when it became known, as was inevitable, that the station was operating in Sweden, the Germans would make trouble, with an unfortunate effect on Anglo-Swedish relations. Hollingworth suggested locating the station in Iceland, but in the end it was based in England in the village of Wavendon near Woburn Abbey.

London, anxious to find out reactions to the clandestine station, told Peter Tennant they believed a Freedom Radio had opened in Denmark, and asked him to find out its effect. He dutifully reported that it was a mobile transmitter having little success. Programmes were 'sometimes good and true, but sometimes quite wrong'. There was the danger that it would lead the Germans to confiscate all radios so that Danes could no longer listen to the BBC and Radio Stockholm. London replied that, as it was impossible to pick up the station in Britain, they would be grateful if Tennant would let them know from time to time what line it was taking.

At this point Turnbull, who unlike Tennant was in the know, put his foot in it by referring to the station in a telegram. The RU angrily demanded that he be 'strafed' for this indiscretion. He was duly admonished in a letter which he was ordered to burn after reading; and 'if you have told anybody in Stockholm that there is any connection in England with this station, will you please contradict it immediately'. A telegram was sent to the Minister stating categorically that the station was in Denmark.

There were in all four Danish clandestine radio stations operated in Britain. *Danmarks Frihedssender* ran until April 1942. It was replaced in July 1942 by *Danish Freedom*, which closed in September 1943. *Radio*

Skagerrak operated concurrently with *Danish Freedom* from February to May 1943; and also with *Hjemmesfronten Radio* which ran from December 1942 to October 1944. With the passage of time *Danmarks Friheds-sender*, as it was christened, became more enterprising. In August 1941 it embarked on a campaign 'to incite the Danes to annoy the Germans by the simple expedient of carrying out German orders to the letter'. Listeners were encouraged to copy the Dutch resistance. When the Nazis decreed that no one should leave a restaurant for fifteen minutes after a German came in, Dutch patriots armed themselves with alarm clocks and set them to ring fifteen minutes after a German arrived. Fifteen minutes later the clocks would go off in unison, and the Dutch would march out *en masse*, within the regulations. The RU hoped the Danes would improve on this. It also decided to localize the RU by giving a disproportionate amount of news from one region, implying that the secret station was there; and, when the Germans began to comb that region, it would concentrate on news from somewhere else.

Nevertheless, it was felt that the Freedom Station was failing to make an impact. Programmes should be interrupted occasionally to suggest Gestapo raids. Much of its material was too academic, the voices were too well educated, there was too much music. The station should seem to be run by a bunch of amateurs in constant danger of their lives. Technical hitches should be introduced. For example, the needle should be placed on the wrong part of a gramophone record. Turnbull thought the programmes were just right, although a few deliberate mistakes would help; but the real reason for the presumed failure of the RU was the attitude of the Danish people. Norwegians had reacted better to their RU because they were

> a people at war, an angry, scrambling, outraged population. Their King, their royal family, and their leaders are all in England, and their eyes are always turned to the west. But the Danes . . . Oh, no! They are a people at peace, an easy-going pleasantly pro-British people at peace, and they will never be budged from this until Denmark is well-bombed and until we force the Germans to take over the government as they have had to do in Norway . . . All our SOE work in Denmark depends on rousing of the Danes' temper and fighting spirit.

Then, said Turnbull, 'you will see fireworks and the RU will be sowing its seed in the same good earth as in Norway', where the Freedom Station seemed to score a bull's eye every time.

The success of the RUs depended largely on the ease with which they were picked up in the target country. Ingenious propaganda is worthless if no one hears it. The Danish RUs continued to be monitored by SOE in Sweden, not necessarily a fair test, as reception varied from country to country. There were long periods when nothing was heard. The monitors failed to pick up a single transmission between 4 and 15 October 1944. Of sixteen transmissions between 22 September and 3 October 1944, thirteen were inaudible. Of the other three, one was audible at first, and then drowned in static. The second, dealing with the discipline of the Danish resistance movement, the impending German defeat, and home events, came over well. The third was also marred by static, but was just audible. It reviewed the situation in Denmark, condemned collaboration, and had a feature on 'the strong and well-organized underground movement'. Twenty-five out of thirty-four transmissions in the month ended 16 September 1944 were inaudible. Nevertheless, the fact that a proportion of the programmes was heard was important. In the grim isolation of occupation the flimsiest contact with friendly voices was a comfort.

In October 1944 Reginald Spink, who had been on loan from SOE for the last year to run the Danish RU, wrote to Stockholm that it had been decided to close it.

> A variety of reasons have gone into this decision, one of them being the increasing difficulty of maintaining originality and freshness . . . No announcement will be made that transmissions will be terminating, and I think it would be best if the facts were not advertised. In all probability the broadcast will break down in the middle and then—the rest will be silence.

Although the Norwegian authorities refused to allow SOE to send in their three-man propaganda teams, they did agree that SOE agents should carry out propaganda work so long as it was aimed at occupying troops, and not designed to influence public opinion. Two missions were sent in for this purpose: *Derby* to the Trondheim area, led by Erik Gjems-Onstad; and *Durham*, which Max Manus and Gregars Gram ran in Oslo concurrently with Operation *Bundle* (see Chapter 5). The work of the propaganda agent in the field is illustrated by the activities of Magnes Nordnes in Trondheim, one of Gjems-Onstad's local recruits. A teacher in the high school, he was originally a *Milorg* courier, but had to sever his connection with the military organization when he was informed against. In May 1944 he took over propaganda in the district when the man in charge, a former pupil of his, had to escape to Sweden. Nordnes avoided recruiting family men, unless they had special qualifications, since the dependants of casualties would be a problem. Most of his group were unemployed graduates. He carried on with his teaching job, which provided useful cover, and used his own first-floor flat in Trondheim as a base. To avoid being surprised by the Gestapo, he never stayed there after 11 p.m., but slept in friends' houses. Most of the propaganda material was kept in the school stationery store, but many group members built secret cupboards in their houses. Through Gjems-Onstad Nordnes requisitioned both black and white propaganda, which was brought by couriers travelling secretly in Sweden and then openly in Norway using their own identity. They carried the material in rucksacks and in specially designed 'kangaroo' pouches inside the courier's trousers.

Nordnes was exceptionally well supplied with black propaganda provided through SOE. During the nine months he operated he distributed 280,000 stickers and 270,000 pamphlets, some of which were produced in Sweden to his own design. One of his major efforts was the production of a black newspaper purporting to be the official organ of the *Deutsche Freiheits Parti* and intended to entice Nazi dissident soldiers into the movement. At first stickers were stuck on shop windows (easy in winter as Trondheim is near the arctic circle, but dangerous in summer when is is never really dark), but when the Germans required shopkeepers to remove them lampposts were substituted. On one occasion stickers appeared on every tram in the town. It was midday before the Germans spotted them and had all the trams taken out of circulation for the windows to be cleaned. On another occasion Victory V stickers were sent to all post offices with a circular signed by the local Nasjonal Samling organization, instructing the staff to display them prominently. The recipients guessed this was the work of the resistance, but they complied with the instruction knowing that the forged circular would save them from reprisals. Nordnes was struck by the fact that the Germans were more sensitive about anti-German propaganda than pro-Allied. A portrait of the King was allowed to remain on a wall in the main street for many weeks; but anything ridiculing or insulting the occupying troops was removed instantly.

Stickers were supplemented with graffiti. In Norway slogans were inscribed with chalk or paint on walls, and traced with sticks in the snow in winter. Among the most popular were: '*Ned med Hitler*' ('Down with Hitler'); the addition of ERLOREN (meaning 'Lost') to the V for Victory sign which the Germans adopted as their own, in an attempt to mitigate its effect; a simple cross cancelling German and quisling proclamations; and on quisling shop windows '*Medlem af NS*' (Member of the Nasjonal Samling). Chalk slogans were not widely used in Denmark because in peacetime they had been a medium for Communist propaganda. However patriotic the slogan, many Danes felt it smacked of Communism. The only exception was the V sign, which had a short life until the Germans took it over.

Black leaflets were distributed to the Germans in a variety of ways. They were slipped surreptitiously into coat pockets on crowded trains, buses, and trams; into rucksacks; thrown over the wire into German camps; left in offices, passages, and public conveniences used by the occupying troops; and hidden in bookshops patronized by them.

Indigenous propaganda was produced on duplicating machines in Trondheim High School, including *For Friheiten*, a resistance newspaper of which 700 copies were distributed once or twice a month. The newspaper was typed by two journalists on Nordnes's own typewriter, as it was impossible to buy a new machine. The keys were bent from time to time to alter the alignment and pressure of the individual letters—a practice that would hardly have deceived expert scrutiny. There was plenty of paper and the limiting factor was the poor quality of the stencils, one of which could reproduce no more than 700 copies. *For Friheiten* was issued to reliable Norwegians who would hand the paper on to friends after they had read it. Each original recipient had his own security number and, if the copy placed in his letterbox—usually at the gate some distance from the front door—did not have that number, he knew it had been planted by the Gestapo, and would leave it undisturbed. The articles for the paper were written locally—BBC news was not included as it was already circulated by word of mouth.

For some reason SOE failed to provide Nordnes with rumours; but he invented his own. They were simple and intended to create alarm and despondency among the German troops. They were never passed directly to the Germans, but reached them through repetition by Norwegians who accepted them as fact. A typical example: when a German ammunition ship exploded in Bergen, the rumour was spread that a nearby explosives dump had been blown up by disgruntled German officers. This circulated widely in the district for two months.

Nordnes devised security rules for his propagandists. They included:

Don't write more than is necessary.
Don't boast about your activities.
Don't hide material behind furniture, inside books, under carpets, in

vases, or behind pictures. These are the first places the Gestapo will search.
Don't get mixed up in open patriotic activity.
Don't speak in riddles—a listener will understand them better than your friend.

Nordnes's group suffered many casualties in spite of its strict security. Fifteen were arrested, twenty had to be sent to Sweden to avoid arrest, and over sixty became compromised and had to abandon propaganda work.

Derby, *Durham*'s sister mission in Oslo, also suffered casualties. Its downfall came through use of a private car which regularly delivered propaganda material to the same addresses—a petrol driven car with a producer gas generator which caught the eye of the Gestapo during a snap control in the evening of 4 April 1945. The driver was forced to disclose the addresses on his round, and led the Gestapo to one, knowing that the occupant would be out. They lay in wait all night, and when two men called next morning immediately opened fire, killing both. The Gestapo then went to another address where their attempt to force an entry disturbed Kolbein Lauring, who was still in bed. When he looked out of the window, he was fired at but not hit. While he was hastily dressing, his wife handed him grenades which he hurled at the Germans below. In the ensuing confusion he jumped from a back window on to the roof of an outhouse and escaped through some gardens. Before the Germans could break down the door, Mrs Lauring telephoned other members of the group to warn them to go underground. She was arrested and subjected to prolonged interrogation, but disclaimed all knowledge of her husband's activities, and was eventually released.

Of Max Manus and Gregars Gram it may be claimed that they were the greatest 'all-rounders' in the resistance movements in Scandinavia, if not in all Europe. Although they had been trained as propagandists in the SOE/PWE school, they were equally well qualified as saboteurs. A typical 'black propaganda' enterprise stage-managed by Manus was the 'leakage' in Oslo of a German folder purporting to come from a dissident German organization. It appealed to German soldiers not to be taken in by propaganda disseminated in army discussion groups, and included an order, ostensibly signed by General Falkenhorst, implying that an Allied invasion of Norway was imminent—perhaps another hint invented by the London Controlling Section, and innocently put across by the propagandists, to back up the idea that Normandy was not to be the Allies' main point of return to Europe.

In addition to propaganda and sabotage, Manus conducted all-embracing enquiries into the state of the Norwegian people. He produced three massive reports summing up his findings in 1943, 1944, and 1945, which SOE and PWE accepted as accurate and unbiased pictures of the true state of affairs. The first of these reports begins by referring to the abortive plan to send in a large number of propaganda agents from Britain:

> A very important aspect of the work which we were to have carried out as propaganda agents would in the first place have been—by means of objective and to some extent scientifically based methods—to transmit to the authorities and the military command in London a picture of the position at home . . . which should be as exact and as reliable as possible. These plans, as is known, were not realized.

Manus and Gram recorded in their first report that they had formed the impression that 'the many accredited ambassadors of the Home Front' who had visited London painted an imperfect picture of the state of affairs in Norway. They believed they were better qualified than most to provide an objective assessment; but they stressed that the reader must bear in mind that they had been in Norway on this occasion for only seven weeks, and that their tasks as saboteurs forced them to live as far as possible undercover. However, between them they had interviewed fifty-nine people from errand boy to shipowner, from bus driver to businessman.

They found that the character of the underground press had changed since they left Norway. Instead of illegal newspapers with editorials, surveys, and propaganda there were now only news-sheets which seemed to be slavish reproductions of the BBC Norwegian broadcasts. Distribution was hand to hand. Most people regarded the news as 'correct and dependable', which Manus could confirm although his own knowledge revealed occasional inaccuracies. The problem was that people got the news in disconnected bits.

> They must themselves deduce the meaning of the news and its tendency. This is all very well in a free democracy but it is not enough for Oslo people under the conditions which we have previously described . . . An obstinate Norwegian—and Norwegians are obstinate—will try to make a piece of news fit in with his views rather than try to let his views be formed by the news.

There were few signs of written propaganda apart from the news-sheets. The aircraft which brought Manus and Gram had dropped leaflets and they met one man who had found some of them. They came across only two examples of indigenous propaganda: one was 'an unreadable 2 pages on Labour conscription written by a helpless amateur' and the other an attack on an unimportant Nazi, 'a thoroughly idiotic, poorly-written screed' probably the work of a 'psychopath who was a personal enemy of the man in question'. Manus was outraged that people should 'willingly risk their lives' carrying this rubbish round to their friends. The BBC could help editors of the illegal press by transmitting daily directives, but it must be a two-way process. London needed to be better informed about the day-to-day spirit in Norway. Specially trained observers should give objective accounts of the situation in the field.[4]

This was attempted in Denmark. Knut Herschend was given a questionnaire—'Where do you think the invasion is coming?', 'Will Denmark go communist after the war?', and so on—which he passed to ten underground editors. They sent it all over the country to people well placed to assess local opinion, including barbers, bartenders and hotel keepers, who replied using code numbers to indicate their status. Up to 1,000 replies were received, which gave a fair reflection of public opinion on which propaganda campaigns could be based.

The clandestine press in Denmark was more highly organized than in Norway. The four main papers were *Frit Danmark*; *De Frie Danske*; *Information*; and *Land og Folk*. The first was duplicated on hand-driven machines by carefully chosen members of large office staffs. A team of six or eight would carry out the work, having posted a guard at the door of the duplicating room. Stencils were not a limiting factor as they were in Trondheim. They, paper, and inks of all colours were in plentiful supply and *Frit Danmark* was distributed by fourteen headquarters to 14,000 recipients every week. There were two pages of straight news and the rest of the paper was taken up with articles on how to resist the Germans, illustrated jokes, and the latest orders from the Freedom Council. *De Frie Danske* was the same size and used the same method of reproduction as *Frit Danmark*. It was a monthly and went to 50,000 recipients throughout the country. At the beginning of the five-day general strike in Copenhagen, 200 helpers produced and distributed 200,000 copies in thirty-two hours—a remarkable feat of journalism even in time of peace. *Information* was a straight news-sheet circulated only to leading members of the underground movement containing important information on internal affairs and external information, for example, news items from the BBC affecting Denmark. Lastly, *Land og Folk* was the only printed newspaper. It was reproduced at a bookshop and often carried illustrations from half-tone plastic stereos sent from London through Sweden. The only printed clandestine newspaper to appear regularly in Norway was produced in Sweden by the Communist Party and sent in by courier.

In Denmark, if news or instructions were too urgent for the periodical clandestine press, they were reproduced specially on posters and distributed in much the same way as the newspapers, within twelve hours of the receipt of the copy. The posters were fixed to the walls and windows, usually by young people.

> Affixation was carried out with extraordinary openness—the distributor frequently riding from wall to wall on a bicycle from whose handlebars depended a briefcase containing the posters and a pot of gum and a brush. For the most part Germans were afraid to interfere with this practice.

While SOE's main responsibility for political warfare was limited to the distribution of propaganda, its Country Sections did take an interest in the type of material it was putting across. It was important that SOE and PWE should see eye to eye. United, they could help each other to attain their objectives. In competition they could—and sometimes did—seriously hinder each other. The method of operation was prescribed by SOE headquarters. Because of the uncertain delivery of propaganda material, missions were expected to keep stocks of the various types—for the underground press, for discussion groups, and for private individuals in key positions who were likely to influence public opinion as a whole. Rumours were to be spread in hairdressers, dressmakers, and similar establishments 'which are hotbeds of gossip and admirably suited for rumour spreading'. An important point was that the rumour must be notified in advance to the intelligence gathering agencies within the country to prevent their reporting as fact a successful rumour—which did occasionally happen. Typical examples of SOE activity on behalf of PWE include the provision of reports from the field in Norway, the establishment of W/T channels, the production in England of miniature radio receivers, the infiltration into Denmark of the miniature edition of the London *Times*, films (for example, *Desert Victory*), and the black booklet instructing German soldiers how to fake illness in order to get their discharge from the army—the last proved very effective, and ultimately led to the death penalty for proven malingerers. SOE also carried in printing equipment for the underground press.[5]

While it was difficult, if not impossible, to measure the effect of propaganda with any accuracy, there was one line which some believed had an adverse effect, especially in Norway. Ivar Naes recorded that, when girls in Alesund went with Germans in the early days of occupation, patriots showed their disapproval through ridicule or silent contempt. As time went on they became more tolerant, which Naes attributed to the 'too placatory tone of Norwegian BBC broadcasts', laying great stress on the importance of lying low and doing nothing. Max Manus, always a stern critic of inaction, said much the same thing: 'The *Milorg* slogan "Wait for The Day" had penetrated into the very blood of the members . . . Waiting can be quite a pleasant form of contribution, whereas all other forms carry with them danger and unpleasantness.' The Norwegian RU was conscious of the difficulty of hitting on a propaganda line which would tell the Norwegians to be patient without giving the impression that nothing was ever going to happen. 'They must save their strength for the day of reckoning'—which never came.

There were mixed views within SOE about the value of political warfare. There was always the danger that the activities of the propagandists, who might be less security-minded than was desirable, would attract the attention of the Gestapo, who could thus be led to other members of the resistance. Headquarters were constantly reminding the saboteurs that they must not become directly involved with the propagandists. According to Turnbull, the latter were 'the froth and bubble on the surface of the work of resistance', the people who in the nature of things were the most exposed and most liable to arrest. He believed that 'the boys' (i.e. the saboteurs) were tempted to try their hands at anything, especially during tedious periods of waiting, 'but at all costs they must avoid being mixed up with this or that glamour boy of the moment who is rushing around talking his head off and exposing himself to inevitable arrest'. Flemming Muus went rather further. In his view, the huge amounts spent on propaganda were totally wasted, and sometimes counter-productive. Leaflets had no effect on the morale of the Germans—they simply made fun of them. A series of propaganda postcards in bad taste had earned the disgust of the Danes who saw them. Muus believed that the only propaganda the Germans understood were bullets and high explosives. They spoke louder than words.

11

D–Day to VE–Day

PLANNING for the Allied return to Europe began in earnest with the appointment of General Eisenhower as Supreme Allied Commander on 24 December 1943. The preliminary work had been in the hands of General Sir Frederick Morgan, COSSAC, who paid close attention to the part the resistance movements would play in support of the invasion, whenever and wherever it came. In March 1944 SFHQ, the division of SHAEF responsible for the direction of the resistance movements (which had taken over the SOE branches concerned with Europe, and the associated OSS personnel), was directed 'to plan and control action to be taken by the resistance groups . . . to conform with the operational plans for a return to Europe'. *Overlord* was to be the supreme operation of the war, and, while the resistance must play its part by stretching German occupation forces to the limit, it must be careful not to put at risk any underground organization whose support the Allies would need round about D-Day.

Since, at least in the early days of the assault on Europe, there would be no land fighting in Scandinavia (unlike in France, where the underground would be called to give direct support to military operations), the task of the resistance would be limited to harassing enemy lines of communication to delay the movement of German reserves to France. But to spell this out to the resistance in Scandinavia would have revealed Allied plans and until well beyond D-Day it had to be left with the impression that it might be called on for direct support to an Allied invasion.

Denmark

In Denmark the remote possibility was briefly allowed to become a probability in the mind of the resistance. SOE had all along stressed the importance of keeping separate the sabotage groups and the secret army, *Chair*, which had replaced the Princes' organization. On 8 April 1944 SHAEF asked for widespread activity in Denmark 'to keep the enemy on the hop'. Muus, who had just sought permission to lay on sabotage for training purposes, was therefore given the go-ahead. He had been told in England that his groups must be 'ready for battle' by March, and he guessed that SHAEF's request for sabotage meant that D-Day was imminent. He activated *Table* and also called out the *Chair* groups. When stocks of explosives were run down to a dangerous level and there was still no sign of invasion, the Freedom Council wanted to know what had gone wrong. On 28 May Turnbull warned London about the consternation over the use of *Chair* in *Table* operations. Hollingworth's defence that only *Table* was supposed to go into action did not hold water, since *Chair* had been specifically instructed to attack *Table* targets. Moreover, the order had been sent by W/T, reserved for *Table*, and not through the BBC, reserved for *Chair*.

Hollingworth dismissed the matter as a storm in a teacup which would subside when the plan agreed at Stockholm was understood by everybody. It didn't matter a scrap. This was astonishing. Muus had committed the cardinal sin of using *Chair* for current sabotage against the wrong targets. Hollingworth's line suggests a guilty conscience over a badly worded order that had failed to make clear what was required and which the Communists saw in a sinister light. It was a device to expose them to the Germans so that they would be liquidated to make life easier for the other political parties post-war. The episode was a serious threat to the unity of the resistance movement at a critical stage in its existence. The Freedom Council's confidence in London was shaken, and it asked for changes in the Stockholm agreement. It should be implemented only in a military situation on The Day and not relate to current sabotage. The agreed evacuation of the M Committee to Sweden need not take place, since no member knew all the details of SFHQ's plans and therefore could not compromise it. Thirdly, it must be consulted about future operations.

It took London some time to grasp that the situation was serious. It was not until 16 June that Hollingworth was compelled to write to all the principal actors trying to pick up the pieces. He confirmed that *Chair* would not be called on again for sabotage. The M Committee need not leave Denmark. The Council would be consulted on major moves, so long as there was no danger to security. The muddle dismissed as being of no importance had marginally strengthened the position of the Freedom Council *vis-à-vis* SFHQ, and marginally weakened the latter's ability to direct the resistance movement according to the requirements of overall strategy. An attempt was made to redress the balance by reminding the Council of its place in the scheme of things. Governments in exile in Britain had separate political and military headquarters. The latter came 100 per cent under the Supreme Allied Commander. Denmark had no government in Britain but must accept the same pattern. Although her political organization, the Freedom Council, had helped to set up the new *Chair* resistance groups, it had no right to interfere with their operations. *Table* and *Chair* were the military organization, with Muus as the chief Allied representative, and they must take their orders direct from SHAEF through the regional commands.

To Muus, Hollingworth said he could not understand what had gone wrong. He, Muus, had been allowed to carry out sabotage for training purposes, which happened to fit in with what SHAEF wanted, and SHAEF, said Hollingworth, had a very good reason for wanting sabotage. He did not reveal it, but it may have been an element in a deception plan to take German eyes off Normandy. It had been wrong to use *Chair* to supplement *Table*; and, while headquarters would of course back up Muus, he must leave *Chair* severely alone and concentrate on building up *Table*.

On the same day he sent Muus SHAEF's Operational Order No. 1 of 14 June, which embodied the Stockholm directive with minor amendments. He stressed that it must be distributed to regional commands without delay and, to rub in the need for speed, he added:

> It is a shameful fact that Denmark, who achieved unity on her fighting front earlier than most other countries, is in fact today lagging behind. Just as I write this I learn that a Division is pulling out of your country probably to reinforce more active battle fronts. SHAEF are calling for delaying action to be taken in all countries along the route it will take . . . thus, because some wise Copenhagen gentlemen are sitting on the BBC code words, the opportunity of harassing this movement will be lost.

He kept headquarters' options open: 'Nobody can tell how soon you will be required to go into action [i.e. the final showdown] but don't jump to conclusions'—as he had done in April.

The Communists had still to be placated. Hollingworth told Mogens Fog how sorry he was about the rumours to which the *Chair* operations had given rise. There was not a word of truth in them. 'I am well acquainted with the splendid work the communists have achieved in the underground movement in Denmark.' As Fog was in close touch with them, he should tell them how much their contribution was appreciated by SHAEF, who supported everyone prepared to fight against the common enemy.

On this day when Hollingworth was doing his best to catch up with events, he sent messages to the regions in Denmark elaborating Operational Order No. 1. SHAEF had strained every nerve to reconcile strategic necessity with local conditions in the occupied countries; but it would always be difficult for the men in the field to grasp the significance of certain orders. Action in Denmark might help the Allies in a battle many miles distant. Orders for widespread sabotage would not necessarily mean that Denmark was on the eve of invasion. 'On the other hand Fighting Denmark may soon be called on to make an all-out effort.' It was repeated that *Chair* was to be used only in a military situation.

Relations with the Freedom Council were now satisfactory; but there were other unknowns in the Danish resistance equation. The leading political parties, Conservatives and Social Democrats, had so far kept and been kept at arm's length from the Council; but now, partly through 'patient and tactful negotiations' by Muus for which London expressed gratitude, the politicians were brought into line. A 'Contact Committee' comprising two members of the Council, a Conservative, and a Social Democrat was formed; and for the first time the overt and clandestine political forces were united.

Secondly, the position of the army had not been clearly defined. Its failure to oppose the Germans in April 1940, and to hinder them after their take-over in August 1943, left doubt about its will to make a contribution at any stage. Again, there was a nagging suspicion that some officers still hankered after playing resistance down until they could take over from disintegrating German forces—a suspicion nurtured by Gyth's line in London (see Chapter 6). Since August 1943 officers who had not escaped to Sweden had formed 'O groups' in parallel with the new *Chair* groups; and their activities must be harmonized with the overall Allied plans. Nordentoft confirmed that 'the Gørtz groups' would co-operate; but Hollingworth thought it necessary to write to Major Schjødt-Eriksen as representing the O groups to remind him that Muus was SFHQ's man in Denmark, and that the main military organization was *Chair*, with the implication that the O groups should not be tempted to go it alone. In July Gørtz confirmed that the Danish General Staff subscribed without reservation to the decisions made at the Stockholm conference. Hollingworth could now breathe freely. He allowed his pen to run away with him: the co-operation of the army was 'the last secure reef knot to be tied round the box which contains the whole of the elements which can be brought into play should a military situation arise in Denmark . . . From now on it can only be plain sailing.'*

At this point there was a domestic interlude. Muus married his secretary Varinka Wichfeldt, whose mother had been arrested for helping the resistance, and sent to a concentration camp in Germany, where she died. On 2 August London sent Muus a message of congratulations concluding: 'Remember underground marriage between illegal people means living in sin . . . Take good care of her.' Mrs *Julep* (Muus's cover name at this point was *Julep*, successor to *Mint*) was told: 'Now we know who is boss in Denmark. Hope you can keep him in order.' When SOE asked for details for War Office records, it was assured everything was above board. They had been legally married under their real names, with *Orange*, *Walnut*, and *Daphne* (Mogens Fog, Herman Dedichen, and Gudrun Zahle) as witnesses. The documents had been buried for safety. In September Muus told Hollingworth that, if the war was not over by January 1945, he would ask for a few months' leave. Invasion was still in his mind. 'I hope you'll see results when D-Day [i.e. D-Day Denmark] comes. We'll not lag far behind the froggies.' Shortly afterwards the newly-weds narrowly escaped arrest by the Gestapo.

There was still one important unknown. If a military situation did arise, who would be the Danish commander-in-chief? Gørtz was tarred with the same brush as the army as a whole, partly because he could have ordered the troops into the field in August 1943, and partly because of his widely publicized presence at the farewell parade of the Danish force recruited by the Germans in 1942 to serve on the eastern front. Although the Freedom Council and army were coming closer together, Muus wanted Gørtz to be given a seat on the Council when Aage Schoch was arrested; but they were unwilling to welcome a man who had been found wanting in the past. Muus was exasperated. He wrote to London: 'These Council people are difficult. Sometimes I feel they do not realize on which side their bread is buttered.' It was not until 11 October that the Council accepted Gørtz as commander-in-chief, with the approval of Vilhelm Buhl, the politician already seen as Denmark's first post-war prime minister. Then, anxious for Denmark to be accepted as a co-belligerent, it asked SHAEF to appoint Gørtz as resistance commander, which was agreed. An announcement would be

* At the beginning of September 1944 the strength of the O groups was put at 7,800 men who could be mobilized in forty-eight hours, compared with *Chair*'s 10,600.

made at the appropriate moment, with due warning so that Gørtz could take precautions, it being assumed that the Germans would make a special effort to liquidate him. Gørtz was reminded that centralization of resistance activities had caused problems in other countries. The regions must be self-contained, taking orders direct from SFHQ. Intelligence work must be kept separate. *Danforce*, the Danish officers and men training in Sweden, should be regarded as a tactical reserve. When he did take over the resistance, he must not surround himself with his own men to the exclusion of experienced underground leaders.

In March 1945 the Council had second thoughts. Prompted by Lippmann, it decided that a more active man was needed. When asked if SHAEF would agree to a replacement, London said only if there were a unanimous recommendation from the Council and a firm assurance that the new candidate had the full confidence of the resistance movement. At the beginning of May the Council asked that the announcement of Gørtz's appointment should be deferred until a full-scale military situation had developed. The resistance trusted its present leaders and to bring in a new man now would cause trouble. If a military situation had arisen Gørtz, or his replacement, would have had very little time to get firmly into the saddle.

A plan to send an Allied force to Denmark at the end of hostilities was abandoned because of the time it would take. Instead there would be a small SHAEF liaison mission under Major General Henry Dewing. When Dewing went to Stockholm in December 1944 to inspect *Danforce*, Hollingworth was filled with alarm. He told Turnbull that, despite strenuous opposition, Dewing had refused to cancel the visit, which SOE feared would be counter-productive:

> Dewing is by no means versed in resistance matters and unless he is carefully led by the hand by yourself he may well blunder in where angels fear to tread . . . we do not want him to concern himself unduly with our organization. It might cause unnecessary difficulties if he becomes sufficiently well-informed to become independent of our advice on affairs . . . We are confident that you will discharge this task having at heart the interests of SOE.

In fact Dewing's visit was a great success. He made an excellent impression during his week in Stockholm. He was favourably impressed by *Danforce*; and he confirmed that there would be no military operations in Denmark, at least until the German high command had been broken.

Hollingworth's confidence that the army would stand shoulder to shoulder with the resistance was shaken by a prolonged feud in Jutland, where the Freedom Council in defiance of the rule to avoid centralization had appointed Major-General Vagn Bennike commander of the important three Jutland regions, over the head of the leader in Region I, the young, very able, and experienced Anton Toldstrup (*Corn*), who saw the appointment as 'the source of much distrust, disquiet, sorrow and despair for us young ones'. He foresaw another sell-out like 9 April 1940 or 29 August 1943 with people like Bennike in authority, 'but of course I am not the King of Denmark'. This was the first shot in a battle that raged for months. London asked Schjødt-Eriksen to mediate, pointing out that 'this show would be nothing without these young people with unbridled fighting spirit'. Regular officers must not monopolize the leadership, and they must abandon old-fashioned military ideas. This reasonable advice caused great offence. The rumour went round that 'someone in London' was gunning for the regular officers; and the fragile nature of the relationship between army and resistance was again highlighted.

Toldstrup foresaw open warfare between resistance and army. Bennike complained that Toldstrup was deliberately undermining him. London assured each that it had the greatest confidence in him, and also in the other. However, Fritz Vang (*Wheat*), the SOE Liaison Officer, had no doubt where the fault lay. The resistance owed everything to the young men, and the old army men were simply qualifying themselves for influential positions. 'I was very disappointed when I saw the General . . . we have not got beyond the situation where the man who wishes to have command can take it by force of previous service, old age, and diplomatic behaviour. Is that what we need?' Further, 'our General in Jutland is opposed to sabotage'. Shades of the Princes. In another report Vang describes visits to the two men.

Spent 2 days with Toldstrup. It was a curious feeling to come from such a slow-working headquarters as Bennike's . . . where one really felt that no one understands the idea of team spirit, where the point is at any price to keep our people [i.e. the civilian resistance] from co-operating; and then to go on to Toldstrup where things happen . . . he has a shining talent for organization and his work is as good and precise as ever illegal work can be. I found him extremely kindly disposed towards working with 'the old chap', or as he is known in North Jutland, 'the flat head'.

London's efforts to settle the row, which was taking up a shocking amount of time, including precious W/T time, were unsuccessful. It thought the trouble was over when Toldstrup was replaced as leader in

Region I to enable him to concentrate on reception, which had been his main activity. Hollingworth told Bennike: 'One final gesture from you in the shape of a sportsman's handshake will eliminate any possible ill feelings.' Alas, Bennike was not the product of an English public school. Instead of shaking hands, he sent Toldstrup an offensive telegram designed to keep the feud alive.

This conflict was much more than a clash of personalities, or an able young man's resentment of subordination to an incompetent elder. Bennike told Johan Malling, the representative at his headquarters of Ole Lippmann (Muus's successor), that when The Day came he could mobilize the army proper, and there would be no need for the resistance. SHAEF was thoroughly alarmed by the implications of this line. On 24 April 1945 Lippmann was sent a telegram in the name of the Supreme Allied Commander instructing him to tell Gørtz that the army's grinding its own axe by refusing to distribute arms was bitterly resented by the resistance who were charged with implementing SHAEF's plans. Whereas the O groups were 70 per cent armed, many civilian groups were going short. It was understood that Bennike, 'basing his plan on a call-up of army personnel', had created large dumps of arms delivered at the cost of many RAF and civilian lives, instead of sharing them equitably among all groups. SHAEF's planners were hamstrung by the failure of officers in Zealand to provide any information about arms there. Equally significant, Bennike had claimed that SHAEF's orders merely indicated the wishes of the Supreme Allied Commander. He was not bound to follow either their letter or spirit, a total negation of the Freedom Council's position that Fighting Denmark was now one of the United Nations.

Lippmann, well aware that the army was forcing the resistance to play second fiddle in the interests of its own position post-war, confronted Buhl, prime minister designate, on 25 April 1945; and, 'after an outburst of anger' (suggesting that Buhl had been siding with the army), he and Gørtz were persuaded to ask the Swedish Prime Minister (Per Albin Hansson) to release the resistance arms accumulated in Sweden and which the army had regarded as theirs. The war ended before these arms could be shipped; but it is interesting to speculate what line the army would have followed had the war been prolonged and the youthful Lippmann failed to make the prime minister designate and the commander-in-chief toe the line.

Even before the Allies landed in France the Danish resistance became engaged in what was little short of open warfare against the

Germans. The *Chair* attacks at the end of April 1944 had led to brutal repressive measures. The Germans set up special courts to try saboteurs and showed no mercy to those found guilty. Patriots already sentenced to death but still in prison were executed in reprisal for sabotage. The Schalburg Corps attacked civilian targets in the hope that the resistance would be blamed. During May there were many executions but sabotage continued. When the Freedom Council said that nothing would deflect the resistance, it was in effect the declaration of a private war—private in the sense that, while it indirectly helped the Allies, its main objective was to hit the occupying power, regardless of the wider struggle. The campaign continued for the whole war, partly in response to orders from London, but largely through the initiative of the indigenous sabotage organizations, BOPA and Holger Danske.

The resistance threw down the gauntlet on 22 June with a major attack on the Rifle Syndicate, an important factory in Copenhagen manufacturing machine guns. Two earlier attacks had led the Germans to introduce special protective measures, and, when the resistance told the factory manager it would leave the place unmolested in return for part of its output, he said it was impregnable. They replied by sending in 125 men dressed as workmen, who disarmed the guards, directed the work people to an air raid shelter, helped themselves to more than 100 light and sub-machine-guns, and planted 400 lb. of explosives at key points. Those parts of the building which survived the subsequent explosions were destroyed by a massive fire which had a firm hold long before the fire brigade arrived; and it was five months before production could be resumed.

The Germans retaliated by executing fifteen hostages and imposing a rigorous curfew. On 26 June workers at the Burmeister & Wain factory in Copenhagen declared they would take a day off every week to compensate for the time the curfew denied them in their allotments. Potatoes were more important than production for the German war effort.[1] Many other workers followed suit and the streets of Copenhagen were filled with angry crowds subjected to indiscriminate attacks by German troops and the Schalburg Corps. Within a week the city was paralysed by a general strike which was so effective the Germans were forced to call off the curfew.

This surprising demonstration of strength by unarmed civilians can only have encouraged the saboteurs, whose campaign went from strength to strength in spite of many arrests and savage reprisals. In September 1944 the Germans began to deport political prisoners to

Germany. They disbanded the police, most of whom had been staunch allies of the resistance. Two thousand were sent to concentration camps where many died, and 5,000 others went underground. The Gestapo became more and more active, and the number of arrests and resistance casualties increased. Captain Larsen, who had succeeded Truelsen as head of the intelligence network which was serving the Allies so well, was arrested in September 1944, as were Aage Schoch and Mogens Fog in October.

The resistance groups' principal contribution to the Allied war effort lay in their attacks on the railway system. On the eve of the Allied landings in France in June 1944, the Jutland resistance groups were told by London that harassing troops moving to reinforce the enemy battle fronts was 'as important as any other future operation'. Although the code action signals had not been circulated to regions at this point, the resistance groups were happily independent enough to go ahead without waiting for the fine-tuned instructions which London had devised; and which, it must be confessed, contained a hint of unrealism not unknown in SOE and SFHQ. It is difficult to see how a group ordered to damage a certain objective so that it would be repairable 'within a week' rather than a fortnight or month could plant its explosives to achieve such an exact result. Most men who were risking their lives in this way would always want to achieve the maximum damage.

In August, September, and October 1944 there were more than 300 acts of sabotage against the railways, which between them delayed eight German divisions on their way to the Western front. The 416th Light Infantry Division *en route* from Himmerland and Vendsyssel should have left Denmark on 6 October but, thanks to the cutting of lines, mining of embankments, and the destruction of bridges and points, was delayed until the twelfth. Many other formations were forced to endure delays at a time when their presence was urgently needed in France. The increased scale of railway sabotage is illustrated by the figures for 1944 and 1945. In the first half of 1944 there were 56 attacks, and 272 in the second half. In the first two months of 1945 there were 247 attacks, 100 of which were in a single week in February. It was in this week that the 233rd Reserve Panzer Division and 166th Infantry Division began to move out of Denmark. So active were the saboteurs that more than half the forty-four trains involved were still held up at the end of the week, and six had been derailed.

While there is no doubt that, from the military point of view, the persistent attacks on the railway system were an important contribution to the Allied effort, it is quite impossible to evaluate them with any degree of accuracy. They might, on the analogy of the missing horse-shoe nail that lost a kingdom, have had a vital influence on battles in France, but it is more realistic to regard them as one of many elements in the final defeat of the Nazis, none of which was in itself conclusive, but which taken together made Allied victory possible. Moreover, the destruction of railway facilities had other benefits. Denmark was an important centre for the training and refitting of German troops, which depended on efficient movement within the country; and Germany relied on supplies of Danish food, much of which was sent by rail.

The importance of railway sabotage was recognized by London, which trained Paul Brandenborg to specialize in this activity (Operation *Cruise*). He was parachuted into Jutland in November 1944, his arrival coinciding with intense Gestapo activity and the arrest of many of the men he had planned to work with. In spite of the handicap of unfamiliarity with local conditions and speaking Danish with a marked American accent (he had spent the last twelve years in the United States), he succeeded in rebuilding the sabotage groups, and himself led more than fifty attacks on the railways. He also turned his attention to a factory making parts for the German V weapons, which was destroyed. In February 1945, at the special request of SHAEF, he put out of commission for the rest of the war the Dars plant at Silkeborg which was making aircraft nose sections. The only casualty in this operation came through failure to obey an order. One of the group was told to arrange for photographs of the damage, but on no account to take them himself. When he could not find the intended photographer, he took a chance and went himself to the scene, where he was arrested. His camera and a gun were enough to seal his fate, and he was later shot.

London had not entirely abandoned the practice of seeing things through rose-coloured spectacles, originally indulged in by Hugh Dalton (see Chapter 3). When a report from Jutland in July 1944 frankly admitted that, although a German division from Norway had been delayed, the effect was 'unfortunately small', the version circulated for the information of SFHQ was careful to omit the field's realistic assessment. The division had been 'successfully delayed' on its journey through Denmark.

The other main target for the saboteurs in Denmark was shipping. In the early part of the occupation the Danish authorities agreed to put their highly developed shipbuilding and marine engine industries at the disposal of the Germans. Under the 'Hansa programme', 100,000 tons would be constructed in Danish yards using materials supplied from Germany. The resistance bided its time while these ships were under construction, carefully monitoring their progress. Then on the eve of launching they attacked them at strategic points, for example, the engine room. So successful were they that none of the first five Hansa ships went into service. In October and November 1944 there were about three attacks each week on shipping in Danish harbours. Typical was the sabotage of the minesweeper *Lenz*, built in Odense. An attack planned by the organized resistance was almost frustrated by 'an amateur group' which carried out an abortive attempt a few days earlier. As a result the Germans posted a heavy guard—five men on the ship, and twenty on the quay. Bork-Anderson, in command, made contact with an electrician putting finishing touches to the minesweeper's electrical fittings, and arranged with him to carry explosives on board in his tool bag the day before the vessel was due to sail. The man placed five incendiary devices at strategic points, including the gunnery tower and electrical control panel; and he lowered a limpet through a porthole on a string. When it attached itself to the hull near the engine room, he threw the string weighted by bolts, so that it disappeared. The fires which broke out were successfully dealt with by the police and fire brigade; but the limpet escaped their notice and blew a hole beneath the waterline, sinking the ship in a matter of minutes. The electrical equipment and engines were destroyed. Although the ship was eventually salvaged, repairs took many months.

Not all shipping sabotage succeeded. Operation *Slide* was an ambitious, carefully planned operation which came to nothing. Three men (Jørgen Christensen, Harald Ryder, and Svend Holm-Hedegaard) were trained as frogmen in England with a view to the sabotage of German ships in Copenhagen harbour, on the basis of a plan devised by Allan Blanner, a Briton working in Denmark who had escaped. The three came on a *Moonshine* trip (see Chapter 4) which reached Lysekil in Sweden at the beginning of October 1944; but, for reasons that are not clear, they were sent to Aarhus—although their training had been based on a model of Copenhagen harbour. When it was proposed to substitute an attack in Aarhus harbour, they readily agreed, but had the misfortune to choose a night when the harbour, usually blacked out, was for some reason brilliantly lit. Regardless of the danger, they swam out to their targets, and planted their limpets successfully on three ships, despite being seen and fired on. Unhappily all three charges failed, probably because the detonators (which they had brought from England with their other equipment), had been damaged by sea water on the voyage to Sweden. After this abortive attempt the three men were kept idle in Aarhus, an example of bad management, or perhaps a deliberate attempt to limit sabotage, which was attributed to the Jutland regional commander, Bennike.

Sometimes the saboteurs had to save themselves by quick thinking. Erik Petersen (*Canada*), in charge of the reception and distribution of arms in Copenhagen, was taking a consignment of Sten guns under a load of hay to a yard for storage, when a few minutes after his arrival the place was surrounded by Germans rounding up labourers working there. *Canada* hastily whispered to the foreman 'to curse him to death' over the poor quality of the hay, which the foreman did so effectively that 'even the Germans were impressed, hearing a Dane curse so well'. *Canada* departed crestfallen, having promised to deliver better quality hay next day.

That Muus may have been running out of steam is suggested by his request in September for leave, if the war had not ended by January 1945. He repeated the request in October, saying that it was really amazing what had been accomplished in the last eighteen months. 'The work has been hard, though, and I know you realise it. I allow me to repeat that if the war is not over by January 1945 I think I shall ask for a holiday—either in Stockholm, or rather in London.' Hollingworth replied that his influence with the Freedom Council would be important for the next few weeks, but he would be welcome in London in December. He certainly deserved a holiday. Of course, if he was seriously compromised he should leave right away. On 17 November Muus reported that the extreme activity of the Gestapo was forcing him to lie low. Forty men were hunting for him, and there was a reward of half a million kroner on his head. Therefore he and *Ulla* (his wife's latest cover name) accepted the invitation to come to London. He recommended that Ole Geisler should deputize for him, which was agreed.

On 12 December Turnbull reported that the pair had safely reached Stockholm, and shortly afterwards Muus flew to London. It was still assumed he would return to Denmark to carry on the fight, and on 3 January it was arranged for him to be briefed on the experience of resistance groups in other parts of Europe. But three days later Hollingworth told Turnbull that Ole Lippmann, who had come to England in the middle of 1944 to be given special training, would go out as a temporary substitute for Muus. Hollingworth wanted Muus to return in due course, in spite of the possible security risks, out of loyalty to an agent who had done well: but rumblings about the chief organizer's financial affairs, which had persisted for a long time, were translated into concrete evidence. Colleagues concerned with resistance finance established that he had received a good deal more than he had paid out. Worse, during his stay in London, Scotland Yard discovered that he was passing five-pound notes, the product not of the Bank of England, but of Nazis. Hollingworth, showing more loyalty than administrative sense, assured Turnbull as late as 17 January that Lippmann was not a replacement for Muus, but merely a temporary relief. He contrived to hush up the affair of the forged five-pound notes (surely a grave error of judgement) but had to accept that Muus should not resume his activities in Denmark. Although it had been widely suspected that the chief organizer was feathering his own nest, the amounts he had purloined, including a 'reserve fund' of Kr. 550,000 deposited in a Copenhagen bank, were not disclosed until after the war. He was found guilty of embezzlement in July 1946 and sent to prison; but his formal punishment must have been as nothing to the opprobrium generated in the minds of his fellows in the resistance. Money is the sinews of an underground movement as much as of war, and the realization that for a large part of the occupation Muus's personal greed had been restricting the movement's means to wage its underground war came as a profound shock to his former comrades.

In Lippmann SFHQ had a first-rate ready-made successor to Muus. A leading figure in the Danish resistance from the earliest days, he had been brought to England to be groomed for a top job in Denmark. By October 1944 he was fully trained and could have been sent home to overlap with Muus before he came out; but it was not until the end of January 1945 that he—Olaf Lund in London, Nils Olgaard as he passed through Sweden, and *Starch* when he finally came to rest in Denmark—arrived with a nice testimonial from Hollingworth. He was 'a good chap' and London expected he would be given all the support in the field that Muus had enjoyed. At first, of course, his identity was known to none, and there was much speculation among the underground about the new boy. In his first letter to headquarters he reported that '*Walnut* [i.e. Hermann Dedichen] has sent out messages calling for reports on the new man—if he is OK, or medium, or bloody awful. I wish I knew, then I would tell him.'

Right up to the moment of his departure, Lippmann was told he would be chief organizer in the absence of Muus, showing that, in spite of everything, Hollingworth still contemplated that the former chief organizer might one day return. While all SOE-trained officers were under the operational control of the regional leaders, who in turn looked to the K Committee of the Freedom Council for their orders, Lippmann was their commander 'for discipline and all matters of policy'—a curious and potentially unsatisfactory relationship, especially if it came to the crunch and the Allies invaded. It was accepted by the K Committee, however, that all reception leaders, whether SOE-trained or not, came under him—reasonable since the provision of arms was SOE's and SFHQ's main contribution to the resistance in Denmark. Lippmann himself was directly responsible to SFHQ. He would have three officers on his staff to be employed as he saw fit.

SHAEF Operation Order No. 2 set the scene for his term of office, which was to last out the war in Europe. It was foreseen that the resistance would find itself in one of five situations. If there was an Allied invasion, it was assumed that the Germans would evacuate forward areas, leaving little scope for resistance groups except for the provision of intelligence. In the rearward areas their main activity must be to harass German reserves. Secondly, if the Germans withdrew but left garrisons, in the ports and bigger towns, the resistance must do its best to interfere with troop movements until the Allies arrived. Thirdly, if an armistice was imminent, the resistance should prepare to receive the surrender. Fourthly, if the central authority in Germany collapsed, it would be up to the resistance to control the enemy forces in Denmark, again until the Allies landed troops. In the case of the fifth alternative—voluntary evacuation—SHAEF was for the time being uncertain what the resistance should do. Hinder, or speed the departing enemy? That would be decided after further consideration.

The order contained instructions for the conduct of resistance, although it might be thought that, after learning the hard way for some years, the Danish underground must know most of the answers. There should be no commitment to pitched battles. Hit and run by small groups must be the order of the day. Never be pinned down. Never do more than eight days' damage to railway track. Dispatch riders, senior and staff officers were important targets for snipers. Road blocks and booby traps should be employed. The objections to central command had been spelled out at the Stockholm meetings:

If operational control were delegated to someone in the field, he would need a master plan on which to base orders. This would give advance information of allied strategy, and under no circumstances could the Supreme Allied Commander allow this. Since then the Supreme Allied Commander had found that centralized control had led to difficulties in other occupied countries, the Danish regions must be self-managing until the resistance commander-in-chief took over. Above all close attention must be paid to orders from SHAEF.

Lippmann found himself in conflict with London as soon as he arrived. He reported that, since Muus had left, everything had been allowed to go badly wrong. 'Intrigues, rumours, nonsense, and rubbish against everything and everyone. You have lost ground to a very great degree, and are not so legendary as in the good old days . . . '[2] On 10 March he wrote to Hollingworth saying that conditions were more difficult than ever. London had lost ground because of failure to take a firm line. 'Please, Holly, please see to it that your chaps don't waste their time on minor personnel questions . . . as long as so many major questions remain unsolved.' The interregnum had in fact seen unfortunate changes. The infant resistance movement of the early years, which SOE had striven to build up, was now a healthy adult, but in need of guidance if it was to give maximum help to the Allies. A strong lead was essential. The Freedom Council was—naturally—becoming more independently minded as the end of hostilities came nearer and post-war problems began to loom larger. The army's independent line (a latent threat from the very beginning), that it was a private underground movement with its own private aspirations, had to be carefully monitored. In particular a strong hand was needed to ensure the equitable sharing of the huge quantities of arms sent in by air and sea. An equally strong hand was called for to prevent the 'Jutland feud' from developing into a much wider conflict, perhaps with the military openly against the resistance.

The new chief organizer did not mince words in his early telegrams. Being seized of the importance of decentralization, he told headquarters that the co-ordination of the three Jutland regions was the biggest blunder ever made. Bennike's appointment as commander was another blunder. Hollingworth yet again found himself defending bad decisions on the ground that one should never admit a mistake. Bennike had been approved because he had the backing of the Freedom Council. SFHQ would continue to accept him as long as he wanted to remain—presumably no matter how great a disaster he might prove. Lippmann on the other hand was reprimanded for putting forward an embarrassing proposal, however sensible. He was told that his frankness was appreciated but his tone deprecated. He must be less impulsive and understand London did not reach conclusions without taking into consideration the overall situation. If Bennike resigned, London would accept any plan agreed between the Freedom Council and Lippmann which conformed to SHAEF's directive to Gørtz; and if it provided for further decentralization, so much the better. Lippmann had the last word in this exchange. He had had a hard time defending London's acceptance of centralization in Jutland; and he resented this charge that he was too impulsive.

Lippmann was of course absolutely right, although London was not prepared to agree. During the interregnum SFHQ lost the chance of making the best use of the resistance forces. Ideally the regional organizations would have taken their orders from London and only London, which would have lessened the danger of snowball arrests. It would have enabled SFHQ to deploy the whole resistance movement in accordance with Allied strategy. Had Lippmann succeeded Muus at the beginning of October 1944 there would have been a chance, perhaps remote, of achieving this ideal command structure.

The impediments which the interregnum allowed to become stronger were the Freedom Council in its own right, and its military wing, the K Committee. As it had been agreed that the K Committee should remain in Denmark, there was one more strand in the command structure to cope with. Again, the O groups, with their threat of independent action, were a menace that could be reduced if they were put firmly under SFHQ. *Chair*, the new secret army which became involved in general sabotage and which was supposed to take orders from London, was also under the control of the K Committee. Fourthly, there were the major sabotage groups, BOPA and Holger Danske, which were virtually self-managing, and a number of smaller groups also their own masters. The only area in which London had sole and positive control—apart from *Table*—was arms reception.

It was necessary to do something about this go-as-you-please situation; and the device adopted was the appointment of SOE-trained Liaison Officers in all regions, with their own W/T channel to SFHQ. These men, very able Captains, were qualified instructors who were put at the disposal of the regional leader, who retained command; and they were able to keep London in close touch with the situation in their region, and do whatever was needed to ensure that Allied strategy was carried out. They would have been of crucial importance had a military situation developed. They were parachuted into Denmark on 30 September and 4 October 1944, and by the time Lippmann arrived were well established; but they did not compensate for the lack of a chief representative who could talk on level terms with the Freedom Council and Gørtz.

If SFHQ was faced with a very difficult task in ensuring that its orders were carried out, it did not help itself by the system of signals devised for the two *Chair* organizations. It was so complex that it was virtually disregarded by the resistance groups, which as a rule were happy to get on with sabotage where and when opportunities presented themselves. The weakness of the system was commented on by the Liaison Officer in the South Jutland Region (Peter Jebsen). For example, there was no guarantee that, when the local *Chair* was instructed to attack trains, a train would present itself at the psychological moment. On three successive nights SFHQ ordered attacks on trains, but no train appeared. 'Our saboteurs found it rather futile to go out on these nights.' It made better sense for a group to find out through its own intelligence network when a train would come, and then attack. Another Liaison Officer observed: 'If it was left to the men they would long ago have stopped all rail traffic in Denmark.'

There is no doubt that SFHQ's greatest contribution during the last phase of the occupation of Denmark was the supply of huge quantities of arms. In the event they did not have to be used in active combat but they were an essential insurance policy lest the Germans put up a last ditch struggle, which presumably SHAEF could not discount. Twenty-six different types of store were sent in through Sweden by air, the principal items being plastic explosive, time pencils, detonators, cordtex, tyrebursters, limpets, S-phones, and miniature receivers. Air dropping was increased in the closing months of the occupation. Whereas in 1942 a mere two packages were dropped, and the monthly average for the twenty-one months from January 1943 to September 1944 was sixteen containers and fewer than two packages, the rates for October 1944 to May 1945 shot up to 695 containers a month and sixty-five packages. The *Moonshine* MTB operation (see Chapter 4) carried 1,046 carbines, 936 Sten guns, 4 Brens, 4 bazookas with 100 rockets, and nearly 2 million rounds of 9 mm ammunition and half a million of .30 and .303. Finally large quantities of weapons were sent from Sweden, evidence that the country now regarded herself as an ally. The last included 3,000 machine guns which were received by the army and distributed primarily to the O groups.

With the safe arrival of these huge quantities of arms, the resistance was equipped to play its part if the German occupation forces refused to go quietly. *Danforce*, which had been gathering strength ever since the first arrival of officers and men in Sweden after 29 August 1943, stood waiting in the wings ready to cross the narrow sound separating Sweden from Zealand in a strangely assorted fleet—seventy fishing cutters, eighty small landing craft, three minesweepers, six steamers, and a ferry boat. The force had been trained to carry out *coup-de-main* operations, street fighting, and guerrilla work, but it was not equal to an opposed landing mainly through lack of the necessary equipment. It would return to Denmark only when the Germans were on the point of surrender.

Even with the German surrender at Luneberg Heath on 4 May 1945, which was supposed to include all German forces in Denmark, there was no guarantee that they would obey the order to lay down their arms. Equally there was no guarantee that the combined forces of the resistance groups and *Danforce* would be able to hold their own against a still active enemy. It was, therefore, planned that a parachute brigade from 21 Army Group would enter Zealand, the rapid advance of the Allied forces having sealed off the Jutland peninsula, and bottled up the Germans there. In the middle of March Hollingworth had visited Stockholm to meet Lippmann to discuss the contribution of the resistance groups to an Allied invasion, if it came to a fight. The plans included the formation of seven Jedburgh teams (two officers and a W/T operator, which had proved very effective in other countries) to spearhead the Allied brigade. On 4 May 21 Army Group ordered *Danforce* to move to Elsinore forthwith to help to maintain law and order; but it now became apparent that the German surrender was going to hold good in Denmark, as it did elsewhere. The resistance came into the open, amid great rejoicing, to do the job of the police force, not yet restored to its normal position.

The suddenness of the surrender and the ceasefire order of 4 May precluded the resistance from implementing plans to apprehend Gestapo officials, war criminals, and Danish collaborators, many of whom made their escape to Germany. Delay by the Allied forces in closing up to the Danish frontier placed the entire burden of border control on the resistance and *Danforce*; and their task was further complicated by the Germans' claim that they were surrendering not to the Danish government, but to 21 Army Group. Here the SOE Liaison Officers, being in uniform, were able to help, acting as it were as the agents of the resistance.

Norway

In August 1944 the young Norwegian lawyer Jens Christian Hauge, who after its initial vicissitudes had built *Milorg* into a force which SHAEF saw as a valuable ally, came to London for discussions with FO IV and SFHQ. Relations between *Milorg* and the Allied high command had steadily improved since the early days when SOE was unaware of the potential of the secret army, and was up to a point competing with the central leadership. Successive meetings with *Milorg*'s leaders had enabled each side better to understand the other's position; and now with Hauge's visit, and a further visit in November, complete accord was reached.*

The various alternatives when the Germans finally threw their hand in were examined and Hauge made it clear that, while the resistance was anxious to get rid of the occupying forces as quickly as possible, it would willingly agree to keep them in Norway, if that was what SHAEF wanted. He had prepared a draft directive to chart *Milorg*'s course for the rest of the war—a massive document seeking to anticipate every possible combination of circumstance, which was accepted with only minor changes. The greatest danger was that the Germans would indulge in an orgy of destruction of key installations and industrial plant which could seriously damage Norway's post-war economy; and since there was no hope of getting an Allied force into the country quickly, it would be up to *Milorg*, now known as the Home Forces, to deal with the situation on its own—a daunting prospect in view of the enemy's numerical superiority. The Home Forces' 40,000 men, 'a quarter of

* One SOE mission (*Antrum*) in Alesund steadfastly refused to recognize *Milorg*'s sole rights. In spite of a proclamation that Crown Prince Olaf was in command of *Milorg*, which left no doubt about its status, Knut Aarsaether told London that the central leadership 'must be made aware once and for all that any interference in our district will not be tolerated . . . they must subordinate themselves to us'. *Milorg* had already done enough harm to the resistance through failure to keep their mouths shut. The local *Milorg* leader (ostensibly absent from Alesund on a skiing holiday) and Aarsaether were summoned to Shetland for 'consultation' which Aarsaether rightly saw as a vote of 'inconfidence'. He was told that his job was to provide information about shipping movements, and help in SOE operations against shipping. Secret army affairs must be left strictly to *Milorg*. (See Chapters 5 and 8.)

them still unarmed, and the rest with light weapons only', faced 365,000 well-armed Germans.

Although Hauge's 'September directive' legislated for other possibilities, including straightforward capitulation, and urged the Home Forces to keep plans flexible, the scorched earth threat was seen as most likely, and most dangerous. It was envisaged that the army of occupation might attempt a heroic last stand, or that individual groups might elect to fight on alone. Later in the year the threat loomed larger when the Russians drove the German forces from Finland into north Norway. The German command evacuated the whole population of Finnmark and Troms, and systematically destroyed every building down to the smallest hut. The 40,000 civilians who had to make their way south as best they could endured terrible privation, and prayed for an Allied invasion to alleviate their suffering.

At the end of December a new directive accorded counter-scorch even higher priority. Experience in the liberated countries confirmed that the Germans would destroy key installations, and suggested that electricity generation, harbours, and telecommunications were likely targets. District leaders must appoint specially qualified men to look after them and be ready to go into action as soon as the enemy pulled out—if necessary without waiting for orders from the central leadership. German troops engaged in demolition should then be attacked with boldness and resolution; but for the time being the resistance must curb any ambition to come into the open and strike at the enemy.

This fitted in with SHAEF's policy of allowing the Germans to withdraw as many men from Norway as they could with the available transport, in the belief that they would be dealt with more easily in the main battle areas. Further, the fewer remaining in Norway, the simpler it would be to overcome them in due course. The policy was reaffirmed on 18 October. The resistance must refrain from attacking communications 'because it is considered inadvisable that the Germans should be hindered in their attempts to leave Norway; and to prevent the Norwegian resistance from being committed to action and becoming exhausted before they are really needed'.

SFHQ did not like this line, feeling that SHAEF was allowing Norway to become something of a poor relation. Scandinavia was under a headquarters concentrating on a major campaign many miles distant, and when that headquarters moved to France there would be even less interest in Norwegian affairs. Gubbins stressed the need for closer concern with Norway. Since there was no prospect of an Allied invasion, the role of special operations was more important. Inaction was having an adverse effect on discipline and morale; and the problem was aggravated by the stream of publicity given to the success of resistance movements in other countries. People felt that Norway had been left out of SHAEF's master plan.

General Sir Andrew Thorne, in command of the force being assembled in Scotland to go to Norway in due course, was one of those who believed that some resistance activity should be encouraged; and on 26 October SHAEF agreed to a limited number of attacks on the railways to keep the resistance on its toes. This was due not so much to pressure from SFHQ as to a change in SHAEF policy, which was now to prevent the movement of fresh divisions to the Continent. Moreover the sinking of the *Tirpitz* by the RAF at Tromsø in the middle of November (see Chapter 5) enhanced this policy since it enabled the navy to take a bolder line against Norwegian coastal shipping, so that, if SOE railway sabotage drove the Germans to send troops by sea, it would provide convenient targets for the navy—and the RAF. The policy made even better sense after the Germans' Ardennes offensive in the middle of December 1944, when SHAEF's confidence that it could easily cope with additional German troops in the field seemed to be misplaced.

It was still considered necessary, however, to avoid committing *Milorg* units, and instead the independent SOE railway sabotage groups which had gone into Norway earlier should be used (see Chapter 9). Although it might seem easy enough for these determined well-trained and well-equipped men to paralyse the railway network in a sparsely populated country, the results were disappointing. *Woodpecker* managed to place a charge in a tunnel near Kongsvoll, but it was triggered off by an insignificant trolley. A later train was damaged when it collided with the wrecked trolley but the damage was slight, and traffic was back to normal within twenty-four hours. The extreme difficulty facing the SOE men is illustrated by the experience of the members of *Fieldfare*, which planned to demolish a railway bridge over the river Rauma. They had to make several trips from a base 30 miles away to carry their 200 lb. of explosives and other equipment to the bridge, struggling through blizzards and snowdrifts, with a daunting climb up to the railway. Their reward for this superhuman effort was an explosion that lifted the railway line a few inches, and slightly damaged one of the arches of the bridge. The Germans made light of the affair. They brought in 100 men from the Todt construction organization who worked in shifts to repair the bridge in less than three weeks. While it was closed, goods were carried across the bridge on sledges from one train to another. According to Wilson's account for headquarters' consumption, traffic was not resumed for three weeks. Rail traffic, yes, but Wilson did not record that sledge traffic compensated for it.

It became obvious that the SOE groups could not make a serious impact on rail movement. Wilson admitted they achieved little. They found it difficult to deal with the heavy guard at most strategic points on the lines, and were often at the mercy of patrols who picked up their ski trails and followed them to their hide-outs. Three *Lapwing* men were killed. It was accepted by SFHQ on 16 January that *Milorg* must be called in to go into action in Districts 14, and 23 to 26 (see Appendix 3), which would not affect its protective activities, since the main targets were all further south.

Milorg had carried out some railway attacks at the end of 1944. The central leadership now carefully planned a multiple operation to delay troops leaving Norway for Denmark and Germany through southern ports. The attacks were carried out simultaneously on 14 March, with over 1,000 men concentrating on four principal lines—Kristiansand to Kongsberg, Eikanger to Drammen, Tinnoset to Brevik, and Kornsjo to Oslo. Ten important bridges were blown, long stretches of permanent way destroyed, and switches at several stations put out of action. The result was serious delays to all rail traffic on these lines for over a month.

These highly successful operations were supported by a daring attack on the headquarters of the Norwegian State Railways in Oslo, by the 'Oslo Gang'. Its leader, Gunnar Sonsteby, gathered round him

> a composite party which could tackle anything. Intelligence, strength, courage, technique, marksmanship, diplomacy, wiliness, forcefulness, patience, devotion, all were represented. Each possessed an almost fanatical loyalty to their King and country.

The members in addition to Sonsteby were Max Manus, Gregers Gram, Birger Rasmussen, Johan Tallaksen, William Houlder, Andreas Aubert, Arthur Pevik, Viggo Axelsen, and Martin Olsen.

The operation against railway headquarters began when the gang duplicated keys to the building. The explosives—180 lb. of dynamite in four boxes—were cached nearby. Sonsteby went in at dusk to remove light bulbs to provide cover of darkness. A few Norwegians were rounded up and taken to safety. The only member of the German guard encountered became 'quite hysterical' and ran at Sonsteby biting and kicking. There was no alternative but to shoot him. Having set only a

4-minute fuse the gang hastily retired to observe the results of their handiwork. The building collapsed like a house of cards. Destruction was complete. The whole of the German guard was killed, but no Norwegians.

SHAEF was grateful for the railway sabotage. It found it difficult to evaluate, but there was no doubt it had helped. By the middle of April it was decided that German divisions in Scandinavia could no longer arrive in time to influence the main battle, and that sabotage policy should again be reviewed. There were three factors in the equation: the degree to which sabotage was forcing the Germans to use coastal shipping; the extent to which land communications in Norway were important for the U-boat campaign; and the need to build the resistance to play an effective part in the final liberation. The last was so important 'that it is believed that Norwegian resistance groups should now be instructed to cease their present activities and build up for the future'. SFHQ took the view that it would no longer discourage the resistance.

The Oslo Gang was, of course, independent of *Milorg* and was not affected by this instruction. The destruction of railway headquarters was one of a long series of daring and successful attacks. In August 1944 Manus had led an operation against the Korsvoll bus depot and a large workshop used to service fighter aircraft. His team got in at 1 a.m. by picking the lock, passed through the main hall where they told the girls washing down the buses that they were black marketeers and warned them to get out. They took in 260 lb. of plastic explosive in suitcase form, and 65 lb. of dynamite in a sack, which they piled in the cellar after cutting their way through a wooden and a wire fence. The charges detonated successfully and all the aircraft were destroyed or seriously damaged. For good measure, six trolley buses fell into the cellar. An equally successful attack was mounted against the Shell oil store in Oslo. The explosives—plastic, dynamite, and Molotov cocktails—were taken in under the noses of the guard which watched with interest as if it was a normal stores delivery. The charges were planted on various floors according to a pre-arranged plan, and petrol was liberally distributed. In the resultant conflagration 1,800 drums of oil badly needed by U-boats were destroyed, and the ruins of the building were still burning next day.

This was part of a general campaign against U-boat activities which was at least as important and as successful as the railway campaign. While the U-boats themselves were quite inaccessible, their supplies, especially oil, were open to attack. An operation against the oil storage depot at Soon in Oslofjord in August 1944 disposed of 700 tons of petrol and 4,000 tons of diesel fuel; and the attacks—about thirty in all—continued for the rest of the war, depriving the enemy of vast quantities of vital fuel. In January 1945 a single-handed attack by a man working in the Horten torpedo store further handicapped the U-boat campaign. Using a 24-hour fuse which acted after 28 hours, he destroyed nearly 200 torpedo heads and '53 tons of charges, as well as stores, barracks, workshops and sundry Germans'. When the armistice came the enemy had only five live torpedoes left in south-east Norway. Also of strategic importance were operations against ball-bearing supplies in response to intelligence that the Germans were about to collect all the bearings they could lay their hands on for dispatch to Germany. *Milorg* carried out three attacks which disposed of all the stocks it could locate.

In April 1945 it was necessary to get hold of police records, which the Germans had already begun to destroy, as evidence against war criminals, and *Milorg* asked the Oslo Gang to help. On a single night it removed more than 2 tons of material from the Department of Justice and Police (including a 1,500-lb. safe from the second floor) in a lorry disguised as a furniture van. These activities greatly helped to sustain the morale of the resistance when it was threatened by war weariness. The Gang thought it owed everything to its leader. 'They had a belief in his invulnerability, and derived a sense of confidence in his ability to achieve the impossible which contributed largely to their success.' Sonsteby himself attributed the success of the Gang to help from a group of State Police (Stapo) constables including Leif Naess, Ottar Mjaerum, Arne Solum, and Per Previk. He wrote:

> It gave me a certain feeling of security as the Gestapo and Stapo could strike at any time and any place, and by having contact with these men I had at all times a certain knowledge of my opponent's movements. I was, for instance, able to learn what was going on amongst the Stapo, who were dangerous and carried out torture, as well as the names of all persons who had been arrested. I knew what sections handled each case and was thereby often able to check for what reason an arrest had been made.

After D-Day Max Manus and Gregers Gram continued their private war on shipping, their first target being the *Monte Rosa*, carrying 3,000 troops and many tanks to Denmark *en route* for Normandy. In Gram's account the pair went openly to the quay 'dressed as plumberers [*sic*] and equipped with tubes and some tools'. They said they were carrying out repairs under the quay, and armed with fake papers had no difficulty in getting in, although they carried a dinghy. They stockpiled their equipment, and waited three long days and nights:

Our nest under the quay would not at all have been so bad if it was not for that we had to share the very restricted space with a legion of rats—numbering at least a few millions. They were of remarkable size, somewhere between a English cat (and that is a big one) and a middle-sized dog. Although it was some nice and hot summer days outside in the at least more free world it was very damp and cold in our nest . . . the dear rats did not make it very pleasant either. As soon as we tried to snatch a short nap they tramped all over our faces. But at least we did develop a technique for shovelling them away while sleeping.

On 19 June the 26,000-ton *Monte Rosa* arrived at one side of the quay, and the 16,000-ton *Moero* at the other—an unexpected bonus. Manus spent the next day preparing the limpets. He later wrote:

It is a queer feeling to sit preparing devilish charges which take the lives of 3,000 human beings . . . many of these soldiers, I felt, were conscripted against their will. Perhaps they were much better men than I was. They had wives and children who loved them above everything on earth.

Then the smaller ship sailed without warning. The pair worked with dangerous haste to fix their limpets to the other. They could see soldiers lining the rail above them but none looked down. Then, while Manus was still at the ship's side, she departed as suddenly as her companion. Left in full view of the people on the quay above, he 'delivered a few not very nice remarks and returned with a fantastic speed'.

Manus guessed the early departure of both ships meant that others were coming in, but decided against waiting. Food was exhausted—and when reports of the *Monte Rosa*'s sinking came, the quay would certainly be searched. So:

We then took our tubes, bade the rats a hearty 'Auf Wiedersehn' and entered the ladder. The very dim light underneath us had made us almost blind and our suits and faces did not make us look very distinguished as we came out in broad sunlight. But we were lucky. There were just a few workers and a German patrolling guard on the quay, no one seemed to take any notice. We passed them without hurrying, showed our papers to the guard at the barrier, and went home for a bath.

They waited for news of the end of the *Monte Rosa*. None came. On 20 July they realized they had failed. Gram recorded: 'I have just got the depressing news that she has again visited Oslo, sound and as fresh as a daisy.' In fact she had been damaged, but only slightly, not enough to put her out of commission.

The *Bundle* team (see Chapter 5) experimented with home-made torpedoes, ingeniously designed after much trial and error. They were used twice. The first exploded short of the target. The second hit the target—a destroyer—but did less damage than was hoped for, although the vessel was out of commission for months, and was thereafter used only as an anti-aircraft ship.

It was limpeting that brought *Bundle* its greatest success in January 1945. Gregers Gram had been killed resisting arrest, and Roy Nielsen was now Manus's lieutenant. They found an empty space at the bottom of a quayside lift shaft where they could store their gear; and, by digging a hole through the concrete 'floor', they could drop into the water and launch a collapsible rubber boat. When they carried in the limpets and boat the guards began to search their van, but when several other vehicles arrived they were waved on. It was now a question of waiting for a suitable ship; and on 15 January the *Donau* docked.

They had still to take in the cordtex—the fuse to detonate the limpets—about 100 yards wound round the waist of each:

> We would hardly be popular if we were searched when we passed the guards. They had had their numbers multiplied manyfold on account of the *Donau*, and were very particular. Roy Nielsen and I arranged a little comedy which worked excellently. It was rather slippery and when we came up to the guard Nielsen skidded and fell on his behind, to the great amusement of the guards. We all laughed at Nielsen (199 centimetres [6 ft. 6 in.] in height) as he lay there sprawling in the street. I laughed so much that I could hardly find the documents. Nielsen cursed. The Germans seemed to think it was great fun and did not examine us carefully. We were now through the first barrier . . .

The pair entered the lift, behaving as naturally as they could. 'When we opened the door anybody standing behind us could look right down on all the limpets and sten guns.' There was also the danger that a guard would shelter in the lift, so an accomplice bolted the door on them. They felt safe, although there was nothing to stop a German from undoing the bolt. They inflated the rubber boat, every stroke of the pump making 'a noise like a trumpet or trombone'. They had eleven limpets, two Sten guns, six hand grenades, and other assorted equipment; and the heavily laden boat might be torn by floating ice. Ten limpets were attached, and then, as they were about to return to their hide-out, another ship, the *Rolandseck*, arrived. They planted the final limpet on her, and returned to the lift. 'After what seemed to be an eternity the blessed contact arrived and let us out.' As they left the harbour the guards recognized Nielsen and warned him to be careful when the going was slippery. '*Donau* tooted goodbye to us, and we took the tram home.'

Towards midnight when the *Donau* was gathering speed, the limpets exploded. The captain ran her ashore at full speed. People jumped out on all sides, on to the shore and into the water. Then she slid backwards and her stern went down in 35 fathoms. She was carrying 1,500 soldiers, including five companies of Alpine troops with full equipment—young, highly trained men who had come from Finland, and several hundred horses and several hundred vehicles.

The Germans immediately examined the *Rolandseck*, and when they found nothing suspicious continued to load her. At 2 a.m. her limpet exploded. The hole was plugged and a salvage boat managed to keep her afloat; but she was so badly damaged she was out of action for the rest of the war.

The last sea operations mounted from Britain were *Salamander II* and *V*, using the new one-man Motorized Submersible Canoes (MSC) popularly known as Sleeping Beauties. Fourteen members of the Linge Company were given temporary commissions as sub-lieutenants RNVR to pilot these craft, and were trained at Staines Reservoir, and Lunna Voe. *Salamander II* established itself at Gangsoy, to attack shipping at Maaloy, but a herd-girl mistook the men for thieving Germans and reported them to the police. They only just escaped and, although they tried to recover their Sleeping Beauties on two consecutive nights, had to abandon them. They were lucky to be able to rendezvous with a sub-chaser from Shetland. *Salamander V* intended to attack shipping in Trondheimfjord but never went into action because extreme phosphorescence illuminated the Sleeping Beauties, towed under water by the *Sylvia*, a fishing boat recently escaped from Hitra. Even submerged at 15 feet the craft shone like silver in bright sunlight. The mission was abandoned. *Sylvia* and Sleeping Beauties were scuttled, and the whole party returned by sub-chaser. No more *Salamander* operations were mounted because the water was now too cold.

After the orgy of railway sabotage in March, *Milorg* was able to concentrate on its main role—the frustration of German scorched earth plans. If it was to succeed, and perhaps also take on individual German units in straight combat, its slender forces had to be armed and equipped; and supplies from Britain were now increased manyfold. The successful invasion of France had released additional aircraft from the Special Duties squadrons, and no effort was spared to drop weapons for distribution in the *Milorg* districts. Between 1 January 1945 and 2 May, when the last flight went in, there were 469 successful sorties delivering 6,850 containers and 1,854 packages, of which 80 per cent safely reached their destination. The arms included 2,250 Bren guns, 3,500 Stens, 17,500 rifles, 2,000 carbines, 50,000 lb. of explosives, and 5 million rounds of ammunition. No less important was the development of the W/T system. It was essential for SFHQ to be in close touch with events in Norway. At the beginning of 1945, there were forty-three W/T stations in contact with London; at the end of March, fifty-nine; and at the armistice, seventy-five. The number of messages exchanged rose dramatically. In 1943 it was a mere 360; in 1944, over 6,500, and in the four months of 1945, 9,123, an annual rate of over 27,000.

The objectives *Milorg* had to protect were listed and circulated to district leaders so that they could make plans well in advance. They included power stations, dams, harbour installations, ships and ship-yards, railways, bridges, telecommunications, and public utilities. Special consideration was given to export industries, on which the country's post-war prosperity would depend. Leaders were asked to collect intelligence about German preparations for demolition, if necessary expanding their existing networks; and to send weekly reports to the central leadership. They were also given suggestions how to operate special master switches which could be fitted in factories to plunge them into darkness when a demolition squad went into action. Telephone lines could be installed to transmit instant news about demolition plans. Vital spares, and drawings of plant and machinery could be hidden so that damage could be quickly repaired. If Norwegians were compelled to carry out demolition work, they should contrive to install dummy or comparatively harmless charges.

Six protective schemes were evolved: *Sunshine*, for vital industrial plants; *Polar Bear*, for harbours; *Foscott*, for the electricity grid; *Carmarthen*, for secondary plants; *Catterick*, worker protection of their own factories; and *Antipodes*, to keep open road and rail communication with Sweden to facilitate the arrival of the Norwegian police battalions training there. The first two, which were organized jointly by SFHQ and *Milorg*, were by far the most important.

Sunshine was designed by Professor Leif Tronstad, who came to Britain in 1942 to advise on the operations against the heavy water plant. He took in a party of eight by air in October 1944 to protect major industrial plants in the Kongsberg and Upper Telemark area, where much of the country's industry was concentrated. There were three related missions: *Moonlight* under Captain J. A. Poulsson (the Rjukan valley); *Starlight* (Lieutenant A. Kjellstrup, the Nummedal valley); and *Lamplight* (Lieutenant H. Nygaard, Tinnoset and Notodden). These missions were responsible for the protection among other important plants of nine major hydroelectric stations. They were under instructions to go into action only on orders from London, or if the enemy actually started demolitions. The Norwegian anti-sabotage guards in all major plants were a help since the resistance could arrange for their own men to plant weapons and ammunition inside. It was also planned to attack German troops near the plants.

Polar Bear was an equally ambitious plan. It was assumed that the Germans might employ the tactics they had used in France where they had destroyed wharves and quays and obstructed harbours with blockships. SOE members of the Royal Norwegian Navy were sent at the beginning of 1945 to thirteen ports to organize an intelligence service to give early warning of any German decision to carry out demolitions. They would also instruct local port officials how to frustrate German plans, for example by replacing demolition charges with dummies, by going slow if they were compelled to plant charges themselves, and by sinking blockships, if they had to, where they would do least harm. The ports covered were: Narvik; Trondheim; Bergen; Stavanger; Kristiansand; Oslofjord; Fredrikstad and Moss; Alesund; Haugesund; Larvik; Sandefjord; and Tonsberg. If Trondheim was anything to go by, the damage might be done before the *Polar Bear* teams went into action. By December 1944 large holes had been dug along the quays and filled with explosives. No leads had been connected, but that could be done at a moment's notice. Landmines had been planted under the bridges leading to the harbour, and all tugs and lighters were prepared for destruction.

Only one *Polar Bear* mission went into positive action. Sub-Lieutenant Inge Stensland, responsible for the protection of Fredrikstad and Moss harbours, entered Norway through Sweden in January 1945, authorized to seize the tugs operating from Fredrikstad, on which the Germans were dependent for moving their shipping in the region. He assembled crews to board eleven tugs, and the salvage vessel *Uredd*, and sailed them to Stromstad in Sweden before the Germans realized what was happening. Not satisfied with this feat, he sent the crews back to Fredrikstad where he arrived in March to put on a repeat performance. By 20 April he was back in Sweden in command of a 650-ton cargo vessel and a small tanker, which it was believed the Germans planned to use as blockships. Again the boarding parties made their way safely back to Norway. Finally, at the end of April, Stensland arranged for the evacuation to Sweden of all the local pilots except one, thereby further paralysing shipping in the difficult waters outside Fredrikstad.

Apart from the rapid build-up of arms, the expansion of the W/T network and the planning for anti-scorch, the main development with *Milorg* after D-Day was the establishment of bases in remote mountain areas, akin to the maquis regions of France, although their origin was different. In the spring of 1944 the Germans planned to conscript 75,000 young Norwegians for forced labour in agriculture, forestry, and building construction, as a result of discussions which Quisling and Terboven had had in Berlin. The plan failed because advance details were published in the underground press to alert the resistance, which destroyed the registration records office and the machines used to print the recruitment documents. Many of the young men affected had time to go underground or escape to Sweden. Fearing a second attempt at conscription, the Norwegian high command and SFHQ selected remote areas in central and southern Norway where the threatened men could safely be hidden. The policy was not to encourage men to flock there—only those seriously at risk would be welcome. They would be given arctic clothing, armed, and trained, so that they could support military operations. SHAEF approved preparations to receive up to 10,000 men but later supplies were limited to enough for 500 men per base. The men were not to go into action without SHAEF approval.

The areas chosen were at Elg in Buskerud; Varg in Vest Agder; Bjorn West in the Nordhordaland Mountains; Bjorn East in Voss-Hardanger; and on the Swedish border at Orm in Hedmark, and in Nørd Trøndelag (see map pp. 30–1). Virtually no progress was made at Orm, Nørd Trøndelag, and Varg. A small group was dropped to the last in November 1944, but, although large numbers of refugees arrived, most stayed only a short time. Four SOE-trained men came by sea from Britain for Bjorn East in December 1944, but the base never got beyond the preparatory stage. Two more SOE-trained men came by sea in October 1944 for Bjorn West, to which 130 men hunted by the Gestapo were sent. At the end of April 1945 a German contingent from Bergen attacked the base for three days. Bjorn lost seven killed, while enemy casualties were put at between 70 and 100. The Norwegians dispersed, but reassembled at the armistice and went to Bergen to take control there. Two men, one a W/T operator, were dropped into the Elg area in August 1944, and by October the base was a going concern. There were courses for group leaders drawn from the surrounding districts. Three more UK-trained instructors came in by air at the end of December. The Germans made several unsuccessful attempts to locate the base but at the end of April 200 mounted an attack. Six Norwegians were killed and the enemy suffered twenty-five casualties, mostly killed.

In September 1944 the SOE Council approved Mission *Scale*, to provide Liaison Officers for the closing stages of the war in Norway. They were to have much the same function as the Regional Liaison Officers in Denmark—to ensure that Allied policy was understood and followed in the *Milorg* Districts, and to keep London informed about the state of affairs in the event of Allied military operations. Five subordinate missions were planned: *Octave*, to be linked with the *Milorg* Elg base, and to include representatives of FO IV and OSS; but the war was virtually over before the mission could be dispatched; *Crotchet* for Bergen, and *Sharp* for Stavanger, suffered the same fate. *Minim*, under Major H. Nyberg, was supposed to carry out liaison with the Norwegian police troops training in Sweden, but failed to get official blessing for this.

The only *Scale* mission to go to Norway was *Quaver*. Led by Major J. C. Adamson, one of the mainstays of SOE's Norwegian Country Section, it was dropped to a reception party sent in from Sweden at Kjernaafjellet in northern Norway, with the special objective of watching the movement of German troops in the area. Because of the difficult terrain the members of the mission were supposed to be dropped at first light, but owing to a miscalculation they arrived in the dropping zone much too late, and were easily spotted. Adamson, who became separated from the rest of the party, damaged an ankle and was picked up by a German patrol. The other three members rendezvoused with the reception committee, but were also spotted and pursued. They managed to escape to Sweden. Adamson was deemed to be a prisoner of war and spent three months in Norway before being transferred to Germany. He was liberated by the Americans from Stalag VIIIa in Bavaria on 30 April 1945.

It was uncertain to the very end how the Germans in Norway would receive the announcement of an armistice. *Milorg*, which had been preparing for four years for just this moment, was left with no clear idea what the Allied high command immediately wanted of it. The many anti-scorch agencies were poised ready to go into action, but there was no sign that their services would be needed.

> On May 7 the Home Forces all over the country were instructed by orders from London to proceed according to the capitulation alternative in the September directive which had been drafted with great caution. Its aim was to prevent clashes. The Home Forces should step forward and take over gradually as the Germans evacuated and withdrew into their reservations.[3]

It had become evident to the Home Front leaders, however, that common sense demanded that *Milorg* should come into the open more quickly in order to meet the fundamental change in the situation. The caution in the September directive could be dangerous if it allowed undisciplined elements in the German forces time to take the law into their own hands. Therefore the *Milorg* leaders ordered their troops to mobilize, a few on 6 May, and most in the following days.

During the night of 7–8 May there were secret contacts between *Milorg*'s leaders and Wehrmacht headquarters at Lillehammer to feel the way forward. The Germans had learned about *Milorg*'s alternative plans from a copy of the September directive which had come into their hands and were satisfied there was no danger of 'a night of the long knives'. They had the good sense to confine their troops to barracks, and there were few clashes. Even so, no formal capitulation had been signed.

At noon on 8 May it was believed the Allies had arrived.

> The Norwegian broadcasting has preserved a recording which gives a lively account: Three yellow cars were appearing. Norwegian police were running ahead of them to make way through the crowd. A tall man was standing upright in the first car with a Norwegian flag in his hand. He was a British general. In the second car you could see a press photographer with a camera. The cars drew up outside the Grand Hotel, hardly able to move through the rejoicing crowd. 'Even the Germans in the street are rejoicing,' says the broadcasting reporter. Shortly afterwards first one and then other Britons appear on a balcony of the Grand Hotel. They make speeches, but it is difficult to hear what they say. Long live Norway, says one in Norwegian. The jubilation has no end. To everybody's surprise the British come out again and travel back the same way they came . . . A rogue of a British Air Marshal had arrived in Denmark . . . The whole thing was irresponsible and sheer madness from beginning to end.[4]

The war correspondent Alan Moorehead has given an account of this practical joke in *Eclipse*.

Tens of thousands filled the streets, 'shouting with joy as if everyone personally was the real victor'. In the principal towns throughout the country

> on the one side the local commander with a tiny bodyguard, in some sort of makeshift uniform, on the other the appropriate high German officer surrounded by the usual galaxy ... The balance of forces was ludicrously disproportionate, but German respect for authority and discipline held good: local agreements were made promptly and respected; and there was no bloodshed and no destruction.

Had there been no secret army to come into the open almost by magic, symbolizing the authority of the Norwegian government, had the Germans believed there was nothing to stop them from looting and pillaging their late enemies, the outcome might have been different. The dedicated patriots of *Milorg* had earned their just reward.

12
Epilogue

Women's contribution

IF it is difficult to do justice to the achievements of all the resistance men in Scandinavia in a single volume, it is impossible adequately to recognize the contribution of the women. In the nature of things they were restricted to unobtrusive backroom jobs, acting as couriers or providing back up for their male counterparts. They did not take part directly in the actions in the field for which the men are remembered, although their secondary role was often vital to the success of the men. In Denmark Edith Bonnesen, shocked by what she had seen of the treatment of the Jews by Nazi Germany, worked for the resistance from the beginning. Using the cover name *Lotte*, she fed the underground press with material obtained through her job in the Ministry of Transport, and gave sanctuary in her flat to men of the resistance, including Svend Hammer, the brother of Mogens. She was arrested three times in the autumn of 1942, but was released after questioning. In spite of the fact that she was thus heavily under suspicion, she extended her work for the resistance, helping Lorens Duus Hansen to find houses from which W/T operators could transmit, and recruiting the many guards needed to protect their activities. She had to go underground after 29 August 1943 when people who had been previously arrested were automatically rearrested, and eventually escaped to Sweden where she continued to do valuable work in connection with *Minestrone* (see Chapter 1).

Aase Parkild was another who early on became involved in resistance work in Denmark. Her house had an attic where a fugitive from the Gestapo could hide, knowing that in an emergency he could make a quick getaway. When her husband was arrested early in 1943 she was subjected to prolonged questioning, but gave nothing away. Her work in the Gentofte registration office allowed her access to ration and identity cards which she purloined for the use of the resistance; and when she was forced to go underground she instructed a colleague how to carry on this invaluable service, and how to manipulate the daily returns so that the losses were covered up. She had several brushes with the police.

On one occasion she was asked where letters from her husband (who had escaped to Sweden) were coming from, as they had Danish stamps and postmarks. She thought it would have been more to the point if they had asked how she addressed her letters to him—but the police were anxious to cut the interview short, as her black cat kept jumping on her, and she guessed they must be superstitious. On another occasion she went to a meeting at a friend's flat. 'The door opened, and there stood 3 men with pistols trained on me. It was the first time I'd been there, so I stood quite still, and thought this was a remarkable reception. I still believe that this saved me, for I must have looked very astonished.' When questioned about her visit, she said she had been invited to a meal of roast pork (which could be smelt in the kitchen) and suggested that, as they seemed to be in for a long session, the policemen should join them for dinner. After four hours she was allowed to leave—and to take with her some of the pork wrapped in newspaper.

Although in Norway *Milorg* mainly recruited men, many women of all ages actively supported the secret army in a variety of roles, especially in communications and the supply services. For example, Mrs Elsa Endresen, widow of a British businessman, made her flat in Oslo available as a safe house for the use of the central leadership, and as a repository for important documents, for nearly the whole of the occupation. She also acted as courier, carrying secret papers in her handbag. The Oslo Gang was also deeply indebted to women. Fru Solveig Wideroe found offices for its members when accommodation was very scarce. She nursed a wounded member of the Gang for three weeks while the Gestapo searched the immediate neighbourhood for him. The Gang was also helped by Fru Gudrun Collet, who provided it with essential food for more than two years.

SOE missions in Nordland were assisted in many ways by courageous women, including the dentist Ingegjerd Hole. In September 1942, when a group had to make a rapid getaway, Ingeborg Palsborg came to its help by driving away a number of mortars which the Germans would have found. Four women of the Grannes family were involved with SOE operations in Nordland. Liv, 23 at the time, was provisionally employed as an extra secretary in the office of the police, where she also handled the issue of special passes. In 1941 she had been working for Odd Sorli, Johnny Pevik, and Nils Uhlin Hansen, who were active for SOE in the Trondheim area, and later became agents of the *Lark* mission there. Liv subsequently became a contact for the *Archer/Heron* Mission in Mosjøen and worked in close touch with its members from their arrival in Nordland at the end of 1941 until the autumn of 1942. She provided them with special passes to allow them to move freely round the area, and arranged for some members to stay with her uncle Erling Grannes at his remote farm on Lake Majavatn. When a German patrol visited the district on a day when *Archer/Heron* was due to send W/T messages to London, and it was feared the Germans might have direction finding equipment, Liv was deputed to go through their packs while they were resting. She found no trace of D/F equipment, so that it was safe for *Archer/Heron*'s transmission to go ahead. She signalled the all clear to the operator by hanging a white towel on a clothes line. There was some amusement among the resistance men when a German soldier who had just had a bath hung his white towel next to it.

Liv's work brought her under Gestapo suspicion, and her telephone was tapped. It was clear that she must leave the country, the plan being that a Catalina due shortly, should take her to Britain from the Island of Trena far out to sea, where she remained hidden. Due to rough weather the Catalina could not land and Liv had to make the long journey to Britain via Sweden, the last leg by aircraft from Stockholm to Leuchars in Scotland. For the rest of the war she worked for the SOE Norwegian Section at headquarters in London.

When *Archer/Heron*'s W/T operator was betrayed, he was compelled to lead a party of Germans to Erling Grannes's farm where four SOE men were living at the time, and where there was an arms dump. Fortunately their two boats were spotted some distance away on the lake, and the resistance men

> at once went into the woods and dug up the arms they had hidden there, and laid an ambush for the Germans, leaving Grannes' family in occupation in the house. The Germans landed and confronted Grannes with the W/T operator, but each denied knowledge of the other. The Germans then made a search of the house and found a wireless set, a few papers, some money and various other odds and ends. They made Grannes and the W/T operator dig in likely spots where the arms might be hidden. At the end of 2 hours they had unearthed 10 rifles and the Germans were very pleased with themselves. They marched back to their boat in a carefree manner, with their guns slung over their shoulders, and with Grannes and the W/T operator carrying the booty behind. When they were 50 yards from the shore the party in the woods opened fire.

Three of the enemy were killed outright, and several others wounded. The survivors took cover behind a boathouse, but the *Archer/Heron* men continued firing through the flimsy wall to give Grannes and the W/T man time to join them.

There was nothing now for it but escape. As soon as the Germans in Mosjøen heard the news of the battle, they would come in force to avenge the death of their comrades. The party made for the west coast near Bindalen where they hoped to find a Shetland Bus boat supposed to be due. They were unlucky and had to retrace their footsteps through very rugged mountain country intensively combed by Gestapo and *Wehrmacht*. The only hope now was to make for the Swedish border. They reached it thanks to superhuman efforts, and Grannes's knowledge of the mountains. 'It was a hard trip . . . the weather was bad, with violent rain and storm, and the food supply and equipment very scanty. The women wore summer dresses and thin shoes, and our men were not much better equipped.' When the Grannes family reached Sweden (the resistance men stayed on in Norway) they were kept in prison for three weeks. They were then flown to Britain where they continued the fight against the Nazis—Erling Grannes in the Royal Norwegian Navy, Dagfrid as an army nurse, Randi and Sigfrid in the Auxiliary Territorial Service (ATS). At this time the Germans uncovered much of the organization built up in haste by *Archer/Heron* in a sensitive area, and their substantial arms dumps. They exacted a terrible revenge. Sixty *Milorg* men were taken in the county and twenty-three of them shot in Trondheim. In reprisal for sabotage further south in October 1942, ten more hostages from the district were shot.

Civilian support

Recognition is also due to the civil resistance in both countries, the men and women who openly defied the wishes of the occupying power, who being 'above ground' were potentially at great risk. When in the course of 1940 the Norwegian Nazi party pressed the professions to accept 'the New Order', they met with universal opposition. A professor of medicine proposed that the Norwegian Medical Association should go along with the new regime, but the national committee rejected the proposal by thirty-five votes to two. Equally, the teachers presented a united front when they were ordered to conform to Nazi thinking and tailor their courses of instruction accordingly. Early in 1941 the Church, distressed among other things by the unlawful activities of the Hird (Quisling's bodyguard modelled on Hitler's SA), joined in the general wave of protest, and in particular supported the stand taken by the teachers, of whom 1,100 were arrested and sent to concentration camp.

Five hundred of the arrested teachers, whom the Germans had tried in vain to cow, were sent in cattle trucks to Trondheim and packed down in the holds of an old coastal steamer, the *Skjaerstad*. After a fearful journey northwards, constantly exposed to German maltreatment and the risk of Allied torpedoes, they were landed at Kirkenes, and set to work loading and unloading and doing other heavy work on a starvation diet. Inside the barbed wire fences the majority were huddled into a cold leaky stable with an earthen floor; later 150 more teachers arrived from the south, who were in the course of the summer housed in cardboard tents.[1]

Only in Trøndelag did some teachers give way. After being subjected to intolerable pressure in prison they agreed to sign the declaration which the Nazi authorities required before they could be freed.

The same spirit manifested itself in Denmark after 29 July 1943 when the population as a whole realized for the first time exactly what Nazi occupation meant. But long before then one civilian group had been dedicated to helping the resistance. From the beginning the doctors had put their professional skills, and the peculiar facilities to which they had access, at the disposal of the underground. It was believed to comprise almost every registered doctor in Denmark which meant that it covered the entire country. They played a particularly important part during the crisis of 29 August 1943, and they were instrumental in evacuating large numbers of Jews in October of that year, when they put their hospitals and ambulances at their disposal in an effort to get them out of the clutches of the Nazis.

In March 1944 Turnbull told London:

> the doctors have shown themselves outstandingly loyal to the saboteurs, and we know that many a saboteur has received treatment and sanctuary in hospitals throughout Denmark. I now refer not only to our own boys and to saboteurs working closely with them, but also to the individual saboteurs or small groups who have worked instinctively, although without co-ordination, against the enemy. The doctors have given their services free to all persons wounded when undertaking sabotage work, and they have helped to hide them and in many cases to evacuate them.

The doctors in Denmark were a grand lot, from Mogens Fog and Chiewitz downwards, and had played a great part in the underground war. So far as he knew there was not a quisling among them, if one named individual, who was reputed to be a horse doctor, was excluded.

Danish intelligence

Although SOE world-wide as a rule gathered intelligence only for its own purposes, in Denmark it exceptionally acted as agent for the SIS. There were two reasons. Denmark was too small a country to have two British agencies working side by side with the danger of compromising each other and competing for facilities such as courier or W/T services. But far more important, the intelligence service of the Danish General Staff was allowed by the Germans to carry on unchanged; and from the beginning of the occupation its reports were made available to the Allies through SOE whose underground connections made it the logical channel. The original arrangements were disrupted by the events of 29 August 1943, but contingency plans had been made and came into operation with very little interruption.

The work of reorganization and running the new service was in the hands of Major S. A. Truelsen, who had been earmarked for it at the beginning of the occupation. Whereas before the German take-over in August 1943 the usual intelligence network had been allowed to function normally, it was now necessary to make arrangements which would be kept secret from the occupying forces. At first material was collected from all over Denmark in the envelopes of an agricultural organization. Since it was in the habit of receiving over 1,000 letters a day, the intelligence communications were something of a needle in a haystack, so that the chances of compromising the system were slight. Nevertheless, since a single letter falling into the wrong hands might blow the whole operation, a new system of poste restante addresses was organized using the services of two postmasters in Copenhagen who had formerly been in intelligence. Letters sent to these addresses were segregated for collection by the intelligence service, and as the addresses were changed at frequent intervals there was little danger that the Germans would get on to the system. An equally secure system made use of the country postmen who were instructed to set aside all letters addressed to a name with an agreed group of initials for delivery to the intelligence organization. A third system involved the delivery of letters addressed to fictitious German hotel guests, which were picked out by the hotel porters and held for collection by the intelligence service courier.

It was necessary after 29 August 1943 to set up a new network to gather intelligence, since there was the danger that former agents had been compromised. Truelsen and his second in command covered the whole country selecting informants well placed to procure information, either because they were working with the Germans or travelling extensively, including doctors, veterinary surgeons, engineers, harbour and railway officials. They were expected to supply details of divisional and regimental numbers, names of senior officers, tactical dispositions, and an assessment of the strength of local forces and any increase or decrease since their last weekly report. The whole country was divided into areas, each of which was looked after by a travelling intelligence officer, who covered the whole of his area every week, usually on a bicycle. The fact that the same men operated in each area meant that continuous contact was maintained with the informants, and that the intelligence officers were themselves well placed to spot changes.

The intelligence officers were given a short training course in Copenhagen before they went into the field. It familiarized them with the working of the system, provided details of the German order of battle generally and in particular in the area where the officer was to work, and covered the technique of acquiring intelligence. There was also instruction in the organization and methods of the *Abwehr* and Gestapo. After the course, which lasted a fortnight, officers were recalled to Copenhagen for a critical examination of their reports, and further instruction in the light of their experience. They were given cover as the representatives of various firms which had agreed to co-operate; and all were provided with false identities. As part of the cover process the firms advertised for additional commercial travellers, with the result that other firms, in the belief that their competitors were seeking new markets, engaged additional staff to meet the competition.

In the early days intelligence material was forwarded via Sweden through the routes used by the other underground organizations, but later the service introduced its own route for specially important intelligence, using boats with specially constructed secret compartments in their water and petrol tanks. So well were these concealed that, when the Germans took over one of the boats, they ran it for many months without ever discovering the secret compartments and their contents. The Swedish defence staff co-operated by arranging that mail addressed to the Danish intelligence service in Stockholm should be sent at once and without examination to the addressee.

Conclusions

The relationship between the SOE and the resistance movements in World War II varied from country to country. There was no standard pattern, nor was there an attempt to achieve one. There were, of course, common factors, including the supply of arms and the infiltration of instructors, but the nature of the help afforded by SOE depended very much on local conditions, geography and politics being by far the most important. In the cases of Norway and Denmark, their political situation and geography could hardly have been more different. While their internal political positions were very similar—both were constitutional monarchies with moderate governments and oppositions—Norway was at war with Germany, Denmark was not. The Norwegian government was in exile in Britain, to which King Haakon VII had come on 9 June 1940 to provide a symbol of national unity. The Danish remained in the country throughout the occupation, more or less willingly collaborating with the Germans until August 1943. The Norwegian high command kept a watchful eye on SOE activities in Norway, and intervened from time to time. For example, it protested strongly against SOE's loan of men from the Linge Company for the combined operations against the Lofoten Islands in the winter of 1941–2, on which they had not been consulted. Later the Norwegian authorities and SOE worked closely together, and a three-way partnership developed when it became possible for *Milorg* leaders, above all Jens Christian Hauge, to come to Britain to discuss future plans. So far as Denmark was concerned, SOE had a much freer hand; or at least SOE did not have to carry a government with it before deciding on a line of action.

In Norway, at war with Germany, there was spontaneous resistance to the occupying power from the very first. In Denmark the bulk of the people followed the lead of their government and accepted occupation, albeit reluctantly. One consequence of this was that many more refugees came to Britain from Norway, so that there was a far bigger pool of Norwegians from which SOE could select men for training as agents. Over the whole war SOE trained more than 650 Norwegians, of whom more than 540 went into the field. Only 150 Danes were trained, of whom about sixty returned to Denmark as agents.

Equally, the geography of the two countries presented very different operational problems. Denmark is small—16,500 square miles—flat, with a relatively evenly spread population of 3.7 million at the time of World War II, although as many as 800,000 were living in the Copenhagen area. Norway is eight times the size of Denmark (125,000 square miles) with a population in 1940 of 3 million, most of whom lived in the south, a quarter of a million of them in Oslo. One important consequence was that it was much easier—relatively—for agents to operate in Denmark, one reason why sabotage played a bigger part there than in Norway.

The fact that the Norwegian government was in exile did give rise to one problem which was absent in Denmark. While the Home Front was absolutely loyal to the King, there were those who felt that the direction of Norwegian resistance should be firmly in the hands of those actually familiar with the conditions of occupation. This feeling was expressed in the report from *Milorg* to the King (referred to in Chapter 8), which said:

> The idea behind our organization is to make available a military machine to support an internal government, secret or official, sanctioned by the King, and acting under direct responsibility to the King. This internal government shall decide if and when the military machine will function. It is of vital importance that this fundamental idea should be approved.

This was a very reasonable proposition from men who were risking their lives and the welfare of their families, to do the government's job for them; and it was equally reasonably rejected by the government. To have agreed would have been to sanction the establishment of something in the nature of a secondary or parallel government. *Milorg* accepted the decision without protest, subject only to the government's recognition of the legality of the secret army. If that had been denied, members of the organization might have found themselves in a precarious position when peace returned. The other side of this coin was the possibility that the government in London might exploit its position to safeguard its own political future at the expense of the legitimate opposition remaining in Norway, an idea the Germans tried to foster in their propaganda; but it never became a serious threat. Any doubts on this matter were removed by proclamations of Christmas 1942 and 9 April 1943 making it clear that, on its return to Norway, the government in exile would place its resignation in the King's hands, clearing the way for an early general election.

A somewhat similar problem, common to both Norway and Denmark, and indeed to all occupied countries, was the danger that the subordination of the resistance to external direction by SHAEF might embroil the Allies in the conflicting aspirations of rival factions in the resistance movement. This problem simmered in Denmark, where the possible contestants were the army, the Communists, and the rest, and it might have boiled over had the war taken a different course, had a military situation developed and the resistance as a whole found itself involved in open warfare with the temptation to believe that the principal victors would be in a strong political position when hostilities ended; but happily it never came to this. In Norway there was never even a hint of using resistance as a springboard to post-war political power. True, the Communists were odd man out also in Norway, often preferring to go their own way in the struggle against the Nazis, but they were far too few to pose any real threat.

> A special and fortunate psychological and political situation existed in Norway. *Milorg* wanted to be part of the Norwegian military establishment, and wanted to be under Norwegian and allied command, and it had established itself as the sole secret military force in Norway. The communists were not able to recruit any following of importance, and in the end were actually included in Milorg.[2]

Quite apart from having to tread carefully through any political minefields laid by competing resistance groups, SOE—and later SFHQ—had to pay heed to the strategic and tactical ideas of the resistance leadership, which might or might not fit in with overall Allied plans. In Norway any possibility of an Allied-organized secret army disappeared when SOE's early attempts to build substantial groups gave way to *Milorg*'s long-term proposals for an underground force directed from Oslo. Similarly SOE abandoned its plans for a secret army in Denmark when the Princes claimed they could do the job better. If it did not actually play second fiddle to the Freedom Council, SOE was compelled to play in a duet (or a trio, if the army is given a separate place on the stage), when a solo might have been operationally more effective; but human nature, and national pride (not a bad attribute in the midst of occupation), can militate against logical perfection and make their effect felt in war no less than in peace. The same might have been true of *Milorg* on the ground that, if a military situation had arisen, the Germans would have had more difficulty in coping with forces directed from Britain whose headquarters would have been invulnerable. Centralization in Oslo was extremely dangerous. Centralization in London was the perfect solution to security problems. But there is no war that hindsight cannot win.

Moreover, close control from London would have been very difficult to achieve. It is easy, still with the benefit of hindsight, to say that, if this mistake had been avoided, the consequences would have been far-reaching, or that, if SOE and the resistance had been enabled to understand each other's position better, a different and more profitable course would have been followed. It must not be forgotten that the underground was underground. Even towards the end of the war when the W/T services were at their most efficient, communication was slow and uncertain. It is difficult enough in battle when the command is in direct touch with its troops to ensure that orders are transmitted efficiently and clearly understood, but when W/T operators are constantly trying to keep one move ahead of enemy direction finders and resistance leaders are forced to think and plan under pressures unknown to the field commander, perfect efficiency is beyond reach. This is something that the student reading the surviving telegrams must not forget.

While it is difficult to assess the contribution of resistance in Scandinavia in isolation, at least it can be put in perspective by examining it in relation to European resistance as a whole. A SHAEF paper of 13 July 1945 to the Supreme Allied Commander, studying the value of the work of the SOE, provides an interesting yardstick for measuring its work in Norway and Denmark. The paper is at pains to stress the point which has been made above (see Preface) that the term 'SOE operations' is not synonymous with 'resistance', which covers 'the amorphous mass of patriots who opposed German rule in the occupied countries'. But without the organization, communications, material, training and leadership provided by SOE (with the help of OSS from November 1943), resistance would have been of no military value.

It is recorded that resistance helped the Supreme Commander's operations in two broad fields: political and military. Politically, organized resistance fulfilled a primary aim of subversion by setting the oppressed peoples at loggerheads with the occupying power. On the one hand, the Germans could not relax, but were kept continually on the qui vive, and were therefore unable to exploit their conquest to the full. On the other hand, the national will to resist was given a focus and an aim. Morale was greatly enhanced by the feeling of support from and contact with the Allies; and, as resistance met with success, national self-respect and confidence were restored, and the desire and ability to resume responsibilities after liberation revived. SHAEF had no doubt that the resistance, if left on its own, could not make a significant contribution to the Allied war effort; but equally a multitude of small aggressive acts could, if properly co-ordinated, produce the military results required by a tactical battle. It had, therefore, been considered essential to keep control of the resistance in Europe in the hands first of COSSAC and then of the Supreme Commander.

SHAEF catalogued the ways in which organized resistance had helped the military operations of the Allied expeditionary force:

1. by sapping the enemy's confidence in his own security and flexibility of internal movement;
2. by diverting enemy troops to internal security duties, and thus keeping them dispersed;
3. by delaying the movement of enemy troops to the Normandy beachhead, and preventing regrouping after the Allied break-out;
4. by disrupting enemy telecommunications;
5. by flank protection and mopping up to enable the Allies to advance with greater speed;
6. by furnishing military intelligence;
7. by providing organized groups in liberated areas to undertake static duties without further training.

Of these seven resistance activities there were two that Norway and Denmark were not called on to undertake—the disruption of telecommunications and support in the field to military operations. They were certainly prepared to deal with telecommunications had they been called on, and they would certainly have given willing support in the field had the Allies invaded their country. In the other five activities both Norway and Denmark played a notable part. Constant sabotage undermined enemy morale and kept much greater forces in Scandinavia, safely away from the Allied invasion, than would have been the case if Norway had become and Denmark remained a model protectorate. Both countries made a material contribution by delaying troop movements. After December 1944 the importance of the German forces in Scandinavia increased since they were one of the few remaining sources of reinforcements, and SOE was directed to hinder evacuation to the maximum. Striking results were obtained in Norway where the policy of forcing troops to travel by sea reduced movement from four divisions a month to one. In Denmark the railway sabotage campaign was no less successful, causing 'strain and embarrassment to the enemy over a considerable period'. Between them the underground movements of the two countries had an undoubted adverse effect on the reinforcement and reforming of units which the enemy had to undertake for the battles both to the west and the east of the Rhine.

So far as intelligence was concerned, the commander-in-chief of 21 Army Group said that, in spite of the fact that the resistance was not trained in intelligence gathering, it had made an invaluable contribution. Denmark was, of course, an important exception, since Truelsen's network was highly trained and highly professional and did a remarkably fine job. The only area where there may have been some doubt in SHAEF's mind about the potential success of the underground in Scandinavia was with regard to its ability to prevent scorched earth. Certainly nowhere were preparations for this activity more carefully made, but SHAEF's experience had been that, where the enemy had time, it was very difficult to prevent demolitions; and in Norway, for example, with the resistance outnumbered nearly ten to one, it seems that nothing could have prevented the enemy from doing very serious damage if it was so minded.

SHAEF pays a generous tribute to the underground in Scandinavia for its ability to put disciplined forces into the field instantly after the surrender, so that when the Allied forces arrived in due course they found the resistance in total command of the situation.

There were, of course, scenes of great emotion in both countries in the days immediately following liberation. On 4 May the announcement of the surrender of all German forces in north-west Germany, Holland, and Denmark was broadcast in the BBC's Danish news bulletins, and in Copenhagen the streets were immediately filled with people able to give expression to feelings pent up for five long years. When the first Allied forces arrived, they were given a tumultuous welcome. General Dewing and Commander Hollingworth drove in state through streets lined with the men of the resistance proudly carrying their arms. Hollingworth later toured the country to meet the men with whom SOE had been in contact for so long, so that he might personally congratulate and thank them. There was no danger of the political upheavals which marred liberation in some occupied countries thanks to an agreement between the politicians and the resistance leaders which divided the first post-war cabinet equally between them. There would be nine politicians, with the Social Democrat Vilhelm Buhl as Prime Minister, supported by three others from his party, two Conservatives, two Farmers' Party, and one Radical. The nine from the resistance were Christmas Møller (who became Foreign Minister) and Henrik Kaufmann representing the Free Danes abroad, Alfred Jensen and Alfred Larsen (the Communists), Arne Sorensen and Juul Christensen (the Danish Unity Party), Niels

Jensen and Frode Jakobsen (the Ring), and Mogens Fog (Free Denmark).

The return to constitutional normality presented no problem in Norway, since the government had continued in office, although in exile, and had in any case undertaken to resign immediately after liberation to make way for a general election. Those members of the Linge Company still in Britain left Aberdeen for Shetland on 13 May and transferred to a sub-chaser for their return to Bergen where they looked after liberated Russian prisoners of war. A group from SOE's Norwegian Section flew to Oslo to be present when King Haakon returned on 7 June—five years to the day since he had left his country. His arrival, on the cruiser HMS *Newcastle*, was greeted with tremendous enthusiasm. The Linge Company lined the way from the jetty to the Town Hall, and among them were the officers of the Norwegian Section. The royal procession was escorted by the members of the Oslo Gang, a well-deserved honour. Max Manus was in the King's car, as he had been in Crown Prince Olaf's when he arrived on 15 May. He wrote later:

The people wept floods and shouted hurrah; but we could hear just one roar, like that of a stormy sea beating against the cliffs. It had taken many years, but now we were there. I thought of my comrades who had given their lives, and of the whole people who had been so brave and faithful. I had long since stopped trying to hold back my tears; no Norwegian had the strength to do that today.[3]

On 9 June 15,000 *Milorg* men from all over southern Norway marched past the King in front of the palace. On the dais was Jens Christian Hauge, who had welded *Milorg* into a force to be reckoned with, and who for the first time in his career as commanding officer was in battledress. A contemporary account records:

The parade was a most inspiring sight. The uniforms and the arms carried were varied in the extreme. The amount and the variety of the weapons astonished the crowd of spectators . . . An interesting feature was the widespread use of code-names as flashes and on helmets – *Bjorn West*, *Polar Bear*, *Lark*, etc.

For the first time the astonishing extent of the underground activity of the last five years was revealed to the man in the street. The Linge Company marched with the men they had helped to train; and the women of the resistance, mostly clad in uniforms fashioned out of parachute silk, were included in the district contingents. The final celebration came on 28 June with a march past of all the irregular units sent into Norway during the occupation from Britain and Sweden. There were 205 men of the Linge Company, and sixty officers and ratings from the sub-chasers. Next day Colonel Wilson inspected the Linge men on their last parade, and bade them farewell on behalf of their British comrades. The Home Forces—as *Milorg* was now known—were stood down on 15 July after two months of valuable service ensuring an orderly return to peace. 'No one centrally or locally had ever had any intention of making political party capital out of *Milorg*. It was truly a National Citizens' Army.'

Primary sources

(other than SOE papers)

Public Record Office

CAB 65	War Cabinet, Minutes
CAB 79	War Cabinet, Chiefs of Staff Committee, Minutes
CAB 80	War Cabinet, Chiefs of Staff Committee, Memoranda
FO 371	General Correspondence, Political
FO 898	Political Warfare Executive

Select bibliography

Ralph Barker, *The Blockade Busters* (Chatto and Windus, 1976).
Jeremy Bennett, *British Broadcasting and the Danish Resistance Movement 1940–1945* (CUP, 1960).
E. Butler, *Amateur Agent* (Harrap, 1963).
Winston S. Churchill, *The Second World War* (Cassell, 1948–54).
E. Cookridge, *Inside SOE* (Barker, 1966).
Charles Cruickshank, *The Fourth Arm: Psychological Warfare 1938–1945* (OUP, 1981).
——, *SOE in the Far East* (OUP, 1986).
T. K. Derry, *The Campaign in Norway* (HMSO, 1952).
John D. Drummond, *But For These Men* (W. H. Allen, 1962).
Facts about Danish Resistance (1965).
M. R. D. Foot, *SOE in France* (HMSO, 1966).
——, *SOE* (BBC, 1984).
Thomas Gallagher, *Assault in Norway* (Macdonald and Janes, 1975).
Tore Gjelsvik, *Norwegian Resistance, 1940–1945* (Hurst and Co., 1979).
Jørgen Haestrup, *Secret Alliance* (Odense University Press, 1977).
E. O. Hauge, *Salt Water Thief* (Duckworth, 1958).
Jens Christian Hauge, *Frigjøringen* (*The Liberation of Norway*) (Gyldendal Norsk Forlag, 2nd edn. 1985).
——, 'Resistance in Norway 1940–1945' (unpublished essay).
Knut Haukelid, *Skis Against the Atom* (William Kimber, 1954).
David Howarth, *The Shetland Bus* (Thomas Nelson, 1951).
——, *We Die Alone* (Collins, 1955).
P. Howarth, *Special Operations* (Routledge, 1955).
Ellic Howe, *The Black Game* (Michael Joseph, 1982).
Peter Kemp, *No Colours No Crest* (Cassell, 1958).
David Lampe, *The Savage Canary* (Cassell, 1957).
Max Manus, *Underwater Saboteur* (William Kimber, 1953).
Alan Moorhead, *Eclipse* (Hamish Hamilton, 1967).
Malcolm Munthe, *Sweet is War* (Duckworth, 1954).
Flemming B. Muus, *The Spark and the Flame* (Museum Press, 1957).
Olaf Reed, *Two Eggs on my Plate* (Allen and Unwin, 1953).
F. Saelen, *None but the Brave* (Souvenir Press, 1955).
Gunnar Sonsteby, *Report from No. 24* (Lyle and Stewart, 1965).
John Oram Thomas, *The Giant Killers* (Michael Joseph, 1975).
Ulf Torell, *Hjalp till Danmark* (Bohuslaningens AB, Stockholm, 1973).

Notes

Introduction

1. M. R. D. Foot, *SOE in France*, p. 18.

Chapter 2

1. FO 371 23663.
2. W. S. Churchill, *The Second World War*, i. 430.
3. Ibid. i. 438.
4. CAB 65 11, f. 6 (2.1.40).
5. *The Times* (2 May 1940).
6. T. K. Derry, *The Campaign in Norway*, p. 110.

Chapter 3

1. CAB 80 56 (25.11.40).
2. CAB 80 63 (27.7.42); CAB 79 22 (8.8.42).

Chapter 4

1. FO 371 29410 (3.41).
2. Ibid.
3. Ibid.
4. Ibid. (26.2.41).
5. Ibid. (22.3.41).
6. Ibid. (27.3.41).
7. FO 371 29425 (28.9.45).
8. Ibid. (3.10.41).
9. Ibid. (29.9.41).
10. Ibid. (25.10.41).

Chapter 6

1. Jørgen Haestrup, *Secret Alliance*, i. 105.
2. *Facts about Danish Resistance*.
3. Ulf Torell, *Hjalp till Danmark*, p. 331.

Chapter 7

1. Haestrup, i. 134.
2. Ibid. ii. 73.

Chapter 8

1. Tore Gjelsvik, *Norwegian Resistance, 1940–1945*, pp. 71–72.
2. J. C. Hauge, 'Resistance in Norway 1940–1945'.
3. Ibid.
4. Ibid.

Chapter 10

1. FO 898 27 (7.7.44).
2. FO 898 57 (10.41).
3. FO 898 51, 57, 73.
4. FO 898 74.
5. FO 898 57, 248.

Chapter 11

1. Haestrup, ii. 326.
2. Ibid. iii. 273–4.
3. J. C. Hauge, *Frigjøringen* (*The Liberation of Norway*), p. 134.
4. Hauge, 'Resistance', p. 130.

Chapter 12

1. Gjelsvik, pp. 30–1, 58–63.
2. Hauge.
3. Max Manus, *Underwater Saboteur*, p. 239.

APPENDIX 1

The Shetland Bus Service 1940–1945

Season	Trips	Passengers Landed	Collected	Refugees picked up	Stores delivered (tons)
1940–1	14	15	18	39	—
1941–2	43	49	6	56	117
1942–3	37	20	2	14	33
1943–4	34	41	13	8	21
1944–5	75	94	33	235	137
Total	203	219	72	352	308

APPENDIX 2

Special Duties operations to Norway 1941–1945

	Year	Sorties Attempted	Sorties Successful	Men	Tonnage of stores	Missing aircraft	Containers	Packages
RAF	1941	—	—	—	—	—	—	—
	1942	22	11	21	6	—	53	14
	1943	50	24	45	24	—	191	95
	1944	193	88	44	141	4	1,207	403
	1945 (Jan.–May)	777	496	65	788	19	7,104	1,599
	Total	1,042	619	175	959	23	8,555	2,111
USAAF	1944	71	41	—	58	1	492	159
	1945 (Jan.–June)	128	57	25	87	4	683	362
	Total	199	98	25	145	5	1,175	521

APPENDIX 3

Special Duties operations to Denmark 1941–1945

	Year	Sorties		Men	Tonnage of stores	Missing aircraft	Containers	Packages
		Attempted	Successful					
RAF	1941	—	—	—	—	—	—	—
	1942	4*	4	12	—	—	—	—
	1943	25	19	21	12	—	100	18
	1944	127	93	17	169	4	1,642	73
	1945 (Jan.–May)	257	168	5	327	12	3,120	290
	Total	413	284	55	508	16†	4,862	381
USAAF	1944	24	9	—	10	1	104	—
	1945 (Jan.–June)	240	125	—	158	—	1,463	232
	Total	264	134	—	168	1	1,567	232

* No record of sorties attempted in 1942.
† No record of aircraft missing in 1942.

APPENDIX 4

Milorg

Norway's secret army, early 1944

	District	Members	Armed	Trained
11	Østfold, part Akershus	3,679	370	3,679
12	Hedmark, part Akershus	2,340	240	few
13	Oslo, part Akershus	5,167	520	5,167
14.1	Parts Buskerud, Vestfold	1,860	620	1,860
14.2	Parts Buskerud, Opland	1,370	1,200	1,370
14.3	Part Opland	1,600	160	1,000
15	Parts Vestfold, Buskerud	1,300	70	200
16	Part Buskerud	400	160	400
17	Telemark (exc. 2 counties)	3,000	1,250	3,000
18	Aust and Vest Agder, part Telemark	2,000	700	1,200
19	Part Rogaland	600	200	few
20.1	Parts Rogaland, Hordaland	1,000	400	700
20.2	Bergen, parts Hordaland, Sogn and Fjordane	3,000	450	1,800
20.3	Part Sogn and Fjordane	800	0	0
21	Møre and Romsdal	200	60	0
22	Sør and Nørd Trøndelag	500	80	0
23	Part Opland	650	600	600
24	Part Hedmark	1,000	400	600
25	Hamar town, part Hedmark	800	170	320
26	Part Hedmark	600	300	300
40	Nordland	850	0	0
41	Troms	—	—	—
Total		32,716	7,950	22,196

APPENDIX 5

Danish resistance, December 1944

Region	Men	Sten guns	Explosives (lb.)
I	1,249	192	4,000
II	3,865	963	6,500
III	1,635	655	5,800
IV	2,000	555	6,600
V	4,700	?	4,400
VI	7,000	50	14,200
Total	20,449	2,415	41,500

Other weapons

Light machine guns	238
Rifles	725
Carbines	5,508
Pistols	2,405
Grenades	5,957
PIATs	7
Bazookas	204

APPENDIX 6

Stores sent to Denmark by air via Sweden 1944–1945

	Plastic explosive (lb.)	Time pencils	Tyre-bursters	Sten guns	Limpets and clams	Detonators	Primers	Cordtex (ft.)	S-phones	Miniature receivers
1944	5,620	1,000	1,000	29	—	1,500	750	1,500	1	—
1945										
Jan.	540	—	—	18	—	2,440	2,370	—	—	—
Feb.	2,340	—	—	—	—	—	—	—	2	—
Mar.	1,777	—	—	—	96	5,550	5,550	8,810	18	31
Apr.	3,301	—	—	—	54	660	1,320	7,490	11	0
Total	13,578	1,000	1,000	47	150	10,150	9,990	17,800	32	31

In addition to the above a wide variety of other stores was sent, including W/T components, photographic equipment, compasses, fountain-pen message containers, etc.

Index